PROCEEDINGS

Optical Fabrication and Testing

Manfred Lorenzen
Duncan R. Campbell
Craig W. Johnson
Chairs/Editors

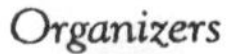

22–27 October 1990
Singapore

Organizers
Institute of Physics, Singapore (IPS)
IEEE Singapore Section
SPIE—The International Society for Optical Engineering

Sponsors
International Commission for Optics (ICO)
Optical Society of America (OSA)

Cosponsors
Schott Glass Singapore Pte. Ltd.
Leica Instrument (S) Pte. Ltd.
Plummer Optics Pte. Ltd.
Loh Optical Machinery Asia Ltd.
Avimo Singapore Ltd.

Published by
SPIE—The International Society for Optical Engineering

SPIE Volume 1400

SPIE (The Society of Photo-Optical Instrumentation Engineers) is a nonprofit society dedicated to the advancement of optical and optoelectronic applied science and technology.

Please use the following format to cite material from this book:
Author(s), "Title of Paper," *Optical Fabrication and Testing,* Manfred Lorenzen, Duncan R. Campbell, Craig W. Johnson, Editors, Proc. SPIE 1400, page numbers (1991).

Library of Congress Catalog Card No. 91-052586
ISBN 0-8194-0483-7

Published by
SPIE—The International Society for Optical Engineering
P.O. Box 10, Bellingham, Washington 98227-0010 USA
Telephone 206/676-3290 (Pacific Time) • Fax 206/647-1445

Printed in the United States of America

OPTICAL FABRICATION AND TESTING

Volume 1400

CONTENTS

(continued)

OPTICAL FABRICATION AND TESTING

Volume 1400

OPTICAL FABRICATION AND TESTING

Volume 1400

CONFERENCE COMMITTEE

Conference Chairs

Manfred Lorenzen, Leica Instrument (S) Pte. Ltd. (Singapore)
Duncan R. Campbell, Avimo Singapore Ltd. (Singapore)
Craig W. Johnson, Plummer Optics Pte. Ltd. (Singapore)

Session Chairs

Session 1—Interferometry
Norman J. Brown, Lawrence Livermore National Laboratory (USA)

Session 2—Interferometry and Associated Techniques
Mike Driver, Avimo Singapore Pte. Ltd. (Singapore)

Session 3—Surface Metrology and Testing
Craig W. Johnson, Plummer Optics Pte. Ltd. (Singapore)

Session 4—Fabrication and Manufacture
Manfred Lorenzen, Leica Instrument (S) Pte. Ltd.

International Advisory Committee

Prof. Dr. Mario Garavaglia (Argentina)
Dr. P. Hariharan (Australia)
Andre Monfils (Beligum)
Alvin Kiel (Brazil)
Prof. Arthur Rajaratnam (Brunei)
Prof. Ralph W. Nicholls (Canada)
Prof. Wang Zhijiang (China)
Prof. Zhang Zhi-Ming (China)
Prof. Eero Byckling (Finland)
Dr. Yuk Tung Ip (Hong Kong)
Dr. G. Lupkovics (Hungary)
Dr. Andrianto Handojo (Indonesia)
Prof. Jumpei Tsujiuchi (Japan)
Yoshito Tsunoda (Japan)
Dr. Roberto Machorro M. (Mexico)
Prof. Richard H. T. Bates (New Zealand)
Dr. Ole J. Lokberg (Norway)
Prof. D. D. Oliverio Soares (Portugal)
Tam Siu Chung (Singapore)
Prof. Sang Soo Lee (South Korea)
Dr. Roland Jacobsson (Sweden)
Prof. Chang Ming Wen (Taiwan)
Supanee Naewchampa (Thailand)

(continued)

OPTICAL FABRICATION AND TESTING

Volume 1400

Dr. Pichet Limsuwan (Thailand)
Dr. Harold E. Bennett (USA)
H. Angus Macleod (USA)
Robert R. Shannon (USA)
Prof. William L. Wolfe (USA)
James C. Wyant (USA)
Professor Hu Ningsheng (China)

Organizing Committee

Manfred Lorenzen, Leica Instrument (S) Pte. Ltd.
M. Haskins, Plummer Optics
Craig Johnson, Plummer Optics
John C. McCallum, National University of Singapore
Don Silva, Polycore Optical
M. H. Kuok, National University of Singapore
S. J. Chua, National University of Singapore
S. C. Tam, Nanyang Technological Institute
T. Li, Optical Society of America
K. L. Tan, National Univeristy of Singapore
D. R. J. Campbell, Avimo Singapore Ltd.
Michael Driver, Avimo Singapore Ltd.
J. C. Dainty, International Commission on Optics

OPTICAL FABRICATION AND TESTING

Volume 1400

INTRODUCTION

The papers in this publication were presented at the Asia Pacific Conference on Optical Technology (APCOT '90) held in Singapore from 22-27 October 1990. The papers are reproduced in their entirety and arranged according to the date and time of the author's presentation at the conference.

The dramatic economic growth of Singapore over the past three decades is well-documented. One of the primary reasons for this growth is the supportive posture of the government toward industries with a high technological base. The optics community in Singapore has flourished in this environment and enjoyed significant growth. APCOT '90 represents a milestone in this growth: several local optical conferences have been held in Singapore, but APCOT '90 was the first conference of an international scale.

Singapore has become an important optics manufacturing center in Asia. As such, it is appropriate that fabrication and testing was chosen as one of the three conference topics. The diversity, timeliness, and quality of the presentations is indicative of the success of the conference.

We would like to express our thanks and appreciation to the organizers and sponsors. We would also like to acknowledge the support of the APCOT '90 secretariat, Conference and Exhibition Management Services Pte. Ltd.

Manfred Lorenzen
Leica Instrument (S) Pte. Ltd. (Singapore)

Duncan R. Campbell
Avimo Singapore Ltd. (Singapore)

Craig W. Johnson
Plummer Optics Pte. Ltd. (Singapore)

SESSION 1

Interferometry

Chair
Norman J. Brown
Lawrence Livermore National Laboratory (USA)

Interferometry with laser diodes

P. Hariharan

CSIRO Division of Applied Physics, Sydney, Australia 2070

ABSTRACT

Laser diodes have the unique property that their output wavelength can be controlled by varying the injection current. This paper discusses some new interferometric techniques that utilise this property.

1. INTRODUCTION

The use of laser diodes as light sources for interferometry is very attractive because of their small size, low power requirements and the ease with which they can be made to operate in a single longitudinal mode. In addition, the output wavelength of a laser diode can be controlled by varying the injection current and can also be modulated over a wide frequency range. These unique properties have opened up many new and interesting possibilities

2. WAVELENGTH MODULATION OF LASER DIODES

A change in the injection current of a laser diode results in a change in the output power due to a shift in the operating point and a change in the output wavelength due to a change in the optical path length of the laser cavity. The second effect can be used to control the output wavelength over a range that is limited essentially by the spacing of successive longitudinal modes, as well as to modulate the output wavelength. Detailed measurements have shown that the variation of the output wavelength with the injection current is due to two causes. One is the change in the refractive index of the active region induced by the change in the density of the charge carriers, which is almost independent of the modulation frequency. The other is the thermally induced change in the refractive index and the length of the cavity, which can be neglected at high modulation frequencies (> 10 MHz), but becomes dominant at low modulation frequencies.[1,2]

Measurements of the changes in the output frequency of a typical AlGaAs double-heterostructure laser diode operating at a nominal output wavelength of 790 nm (corresponding to an output frequency ν = 379 THz) for injection currents i ranging from 49 mA (threshold) to 69 mA, showed that at modulation frequencies below 200 Hz the tuning rate ($d\nu$ / di) was 2.3 GHz/mA.[3] The spacing between successive longitudinal modes was 143 GHz. It was found that large variations in the injection current sometimes resulted in mode jumps. However, with a proper choice of initial operating conditions, a linear relation between the output frequency and the injection current, free from discontinuities due to mode jumps, was obtained for a variation in the current of 10 mA, corresponding to a tuning range of 25 GHz

3. FREQUENCY-MODULATION INTERFEROMETRY

A problem in interferometric measurements of distances greater than a wavelength is that the integer order of interference cannot be determined directly from the fringes. Normally, measurements with two or three wavelengths are required to obtain the integer order. Since laser diodes can be tuned electrically over a range of wavelengths, direct measurements of long optical paths are now possible.

One method involves measurements of the change in the phase difference produced by a known shift in the wavelength.[4] If L is the optical path difference between the beams in an interferometer and the laser wavelength is changed from λ to $(\lambda + \Delta\lambda)$, where $\Delta\lambda \ll \lambda$, the phase difference between the beams changes by an amount

$$\Delta\varphi = 2\pi L\left[\frac{1}{\lambda} - \frac{1}{\lambda + \Delta\lambda}\right] \approx 2\pi L\Delta\lambda/\lambda^2 \tag{1}$$

A setup using this method is shown in Fig. 1. The phase change $\Delta\varphi$ can

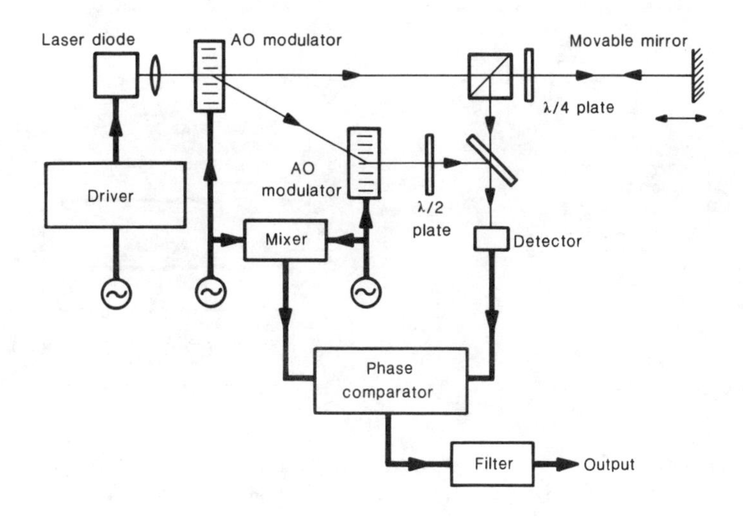

Fig. 1. Arrangement for frequency modulation interferometry using a tunable laser diode (Kikuta, Iwata and Nagata, 1986).

be measured accurately by heterodyne techniques. A fixed frequency difference is introduced between the two beams in the interferometer by a pair of acousto-optic modulators operated at slightly different frequencies. If the electric fields corresponding to the two beams emerging from the interferometer are represented by the relations $E_1(t) = a_1 \cos(2\pi\nu_1 t + \varphi_1)$ and $E_2(t) = a_2 \cos(2\pi\nu_2 t + \varphi_2)$, where a_1 and a_1 are the amplitudes, ν_1 and ν_2 the frequencies, and φ_1 and φ_2 the phases of the two waves, the output from the detector can be written as

$$I(t) = \tfrac{1}{2}a_1^2 + \tfrac{1}{2}a_2^2 + \tfrac{1}{2}[a_1^2 \cos(4\pi\nu_1 t + \varphi_1) + a_2^2 \cos(4\pi\nu_2 t + \varphi_2)]$$
$$+ a_1a_2 \cos[2\pi(\nu_1 + \nu_2)t + (\varphi_1 + \varphi_2)]$$
$$+ a_1a_2 \cos[2\pi(\nu_1 - \nu_2)t + (\varphi_1 - \varphi_2)] \quad (2)$$

The output from the detector contains an oscillatory component at the difference frequency $(\nu_1 - \nu_2)$. The phase difference $(\varphi_1 - \varphi_2)$ between the two interfering waves at any selected point can then be determined by comparing the phase of the electrical signal from the detector with that of a reference signal.[5] Accordingly, if λ and $\Delta\lambda$ are known, the value of L can be obtained.

Direct measurements of distances can also be made by a technique analogous to FM radar.[6] A system that combines heterodyne techniques with fringe counting to measure both relative displacements and absolute distances with high accuracy is shown schematically in Fig. 2.

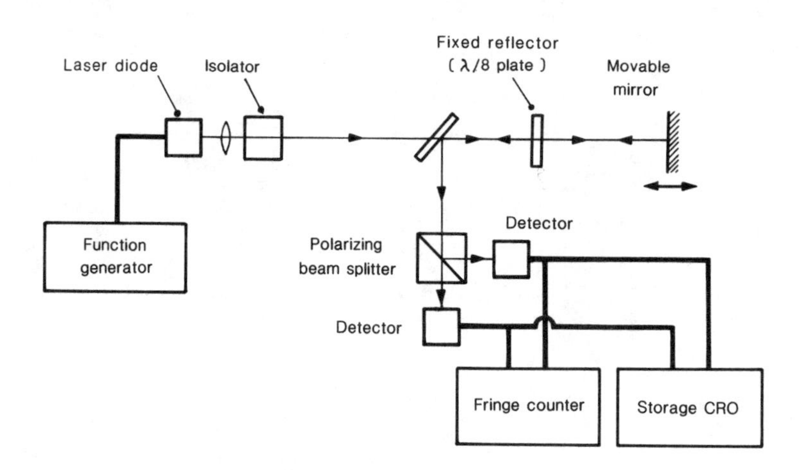

Fig. 2. Fringe-counting interferometer using a tunable laser diode (Kubota, Nara and Yoshino, 1987).

Interference takes place between a reference beam reflected from the front surface of a $\lambda/8$ plate and the signal beam reflected from a movable mirror. Since the signal beam passes twice through the $\lambda/8$ plate, it returns as a circularly polarized beam. The two orthogonally polarized components of this signal beam are divided at a polarizing beam splitter to produce two fringe patterns whose intensities vary in quadrature. Accordingly, the sign of a displacement of the mirror as well as its magnitude can be determined by a fringe counter.

For absolute distance measurements the frequency of the laser is swept linearly with time by using a function generator to vary the injection current. For an optical path difference L, the two beams reach the detector with a time delay L/c, where c is the speed of light, and they interfere to yield a beat signal with a frequency $\Delta\nu = (L/c)(d\nu/dt)$, where $d\nu/dt$ is the rate at which the laser frequency is varying with time. Distances of a few metres can be measured with an accuracy of 100 μm.

4. PHASE-LOCKED INTERFEROMETRY

The intensity at the output of an interferometer is a periodic function of the phase difference between the two interfering beams which, in turn, is determined by the optical path difference and the wavelength used to illuminate the interferometer. Laser diodes can be used to set up phase-locked interferometers in which a change in the optical path difference is compensated by a change in the illuminating wavelength.

In the simplest arrangement, the output intensity from the interferometer is compared with a reference intensity, and the difference signal is amplified and used to control the injection current of a laser diode and hence its output wavelength. If the gain of the feedback loop is sufficiently large, a high degree of stabilization can be achieved, and the injection current is a linear function of the displacement.[7]

Better stabilization can be achieved, and dc drifts can be eliminated, with a system in which the injection current of the laser diode is modulated sinusoidally with a small amplitude.[8] In the arrangement shown in Fig. 3, the injection current of the laser diode consists of a dc bias current i_0, a control current $i_c(x)$, and a sinusoidal modulation current $i_m(t) = i_m \cos \omega t$, so that the phase difference between the beams in the interferometer is subjected to a sinusoidal modulation $\Delta\varphi(t) = \Delta\varphi \cos \omega t$ with an amplitude $\Delta\varphi \ll \pi$. The intensity at any point in the interference pattern can then be written as

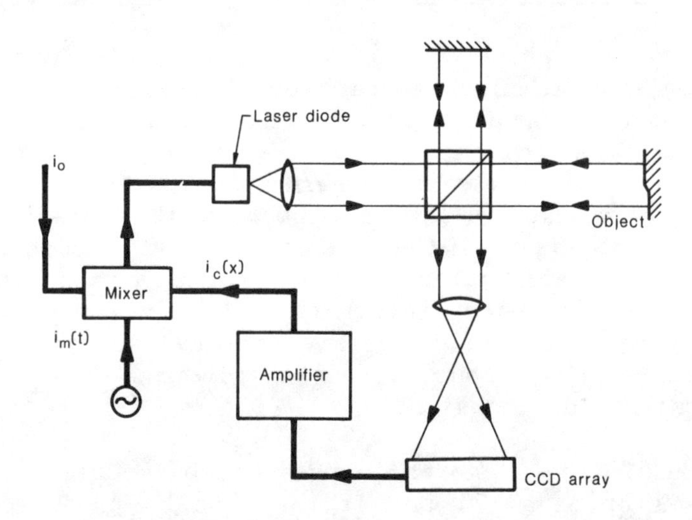

Fig.3. Phase locked laser diode interferometer for surface profile measurements (Suzuki, Sasaki and Maruyama, 1989).

$$I(t) = I_1 + I_2 + 2(I_1 I_2)^{\frac{1}{2}} \cos [\varphi + \Delta\varphi(t)] \tag{3}$$

where I_1 and I_2 are the irradiances due to the two beams taken separately, and φ is the mean phase difference between them. The last term on the right hand side of Eq. (3) can then be expanded as an infinite series of Bessel functions to give the result

$$I(t) = I_1 + I_2 + 2(I_1 I_2)^{\frac{1}{2}} [J_0(\Delta\varphi) + 2J_2(\Delta\varphi)\cos 2\omega t + \ldots] \cos \varphi$$
$$-2(I_1 I_2)^{\frac{1}{2}} [2J_1(\Delta\varphi) \sin \omega t + 2J_3(\Delta\varphi) \sin 3\omega t + \ldots] \sin \varphi \tag{4}$$

Since $\sin \varphi = 0$ when $\varphi = m\pi$, where m is an integer, the amplitude of the signal at the modulation frequency drops to zero at these values of φ. Away from these positions, a signal is obtained with an amplitude $4(I_1 I_2)^{\frac{1}{2}} J_1(\Delta\varphi) \sin \varphi$. Since both the magnitude and the sign of this signal change as it passes through zero, it can be used as the input to a servo system that varies the control current $i_c(x)$ so as to lock the phase at this point. Changes in the optical path can be derived from measurements of the control current.

A convenient way of making measurements across an object is to use a linear CCD sensor as the detector.[9] A sample-and-hold circuit picks up the signal corresponding to any specified measurement point from the output of the CCD. If the period of charge storage in the CCD is much smaller than the period of the phase modulation, a continuous signal can be obtained for each measurement point. Surface profiles of rough surfaces can be plotted with an accuracy of a few nanometres.

5. SINUSOIDAL PHASE-MODULATING INTERFEROMETRY

In this method the phase of one beam in the interferometer is modulated with a larger amplitude than that used for phase locking (typically around π radians).[10] The modulation amplitude can then be determined from the ratio of the amplitudes of the components in the output of the detector at the modulation frequency and at its third harmonic. Once the modulation amplitude is known, the average phase difference between the interfering beams can be determined from a comparison of the amplitudes of the components in the output of the detector at the modulation frequency and at its second harmonic. Data processing can be speeded up by integrating the time-varying intensity in the interference pattern over four successive intervals, each corresponding to a quarter period of the phase modulation.[11] This procedure eliminates the need for frequency analysis of the signal.

The phase-modulation is normally produced by mounting one of the mirrors of the interferometer on a PZT but it can also be produced very effectively with a frequency-modulated laser diode.[12,13] An interferometer using such an arrangement is shown schematically in Fig.4. Since modulation of the

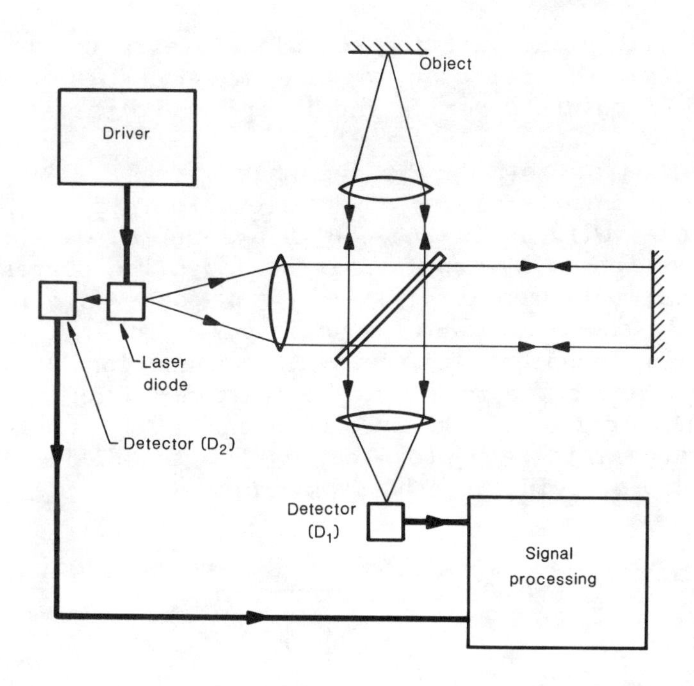

Fig. 4. Sinusoidal phase-modulating interferometer using a tunable laser diode (den Boef, 1987).

injection current results in an intensity modulation as well as a frequency modulation, this intensity modulation is eliminated by electronically dividing the output from the interferometer by the signal from a reference detector viewing the laser diode directly.

6. PHASE-SHIFTING INTERFEROMETRY

In this method the optical path difference between the interfering wavefronts is changed in a number of equal steps and the corresponding values of the intensity at each data point in the interference pattern are measured and stored. The intensity values at each point can then be represented by a Fourier series whose coefficients can be evaluated to obtain the original phase difference between the interfering wavefronts at this point.[14] A simpler version of this method involves four measurements at each point, corresponding to three equal phase steps. These provide enough data to calculate the original phase difference between the interfering beams as well as the phase step. If the phase step is known, only three measurements of the intensity are required.

Alternatively, the optical path difference between the interfering beams is varied linearly with time, and the output from a detector located at a point on the fringe pattern is integrated over a number of equal segments (typically four) covering one period of the sinusoidal output signal. From the standpoint of theory, the two methods are equivalent.[15]

An advantage with phase-shifting techniques is that a two-dimensional CCD array can be used as the detector to make measurements simultaneously at a very large number of points covering the interference pattern.

The usual method of introducing the phase steps is by mounting one of the mirrors in the interferometer on a piezoelectric transducer (PZT) and applying appropriate voltages to it. With a laser diode the phase steps can be introduced conveniently by modulating the injection current so as to shift the output wavelength appropriately.[16,17] A phase-shifting interferometer using a laser diode source is shown schematically in Fig. 5. An interesting application of this technique has been in conjunction with a radial shear interferometer to measure the wavefront aberrations of an optical disk head.[16] For tests in the optical workshop, wavelength shifting is particularly attractive with large interferometers because of the difficulties involved in using PZTs for phase stepping with heavy mirrors.

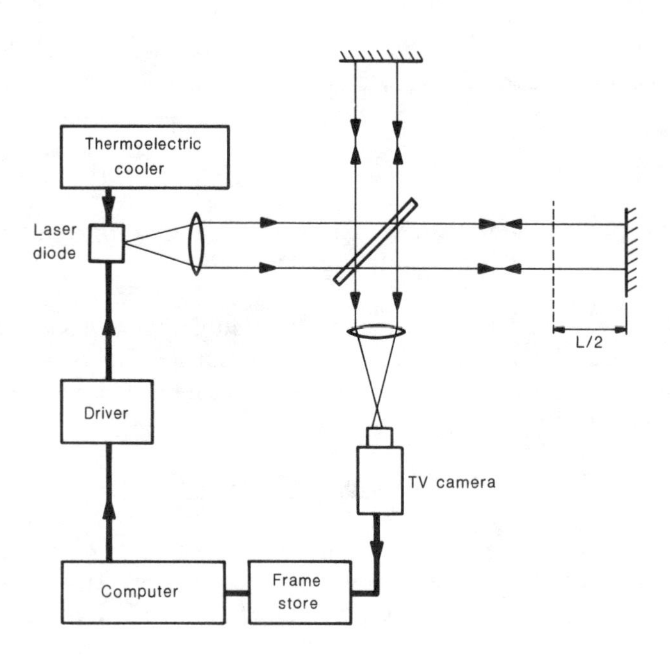

Fig. 5. Phase-shifting interferometer using a tunable laser diode (Ishii, Chen and Murata, 1987).

Accurate measurements require the laser wavelength to be stabilized during each measurement. Since the change in output frequency with temperature is, typically, about 35 GHz/K, a major cause of variations in the output frequency is temperature fluctuations. These can be minimized by stabilizing the laser temperature to ± 0.001 K by means of a thermoelectric cooler. More precise stabilization requires the laser wavelength to be locked to a resonance of a reference cavity (a Fabry-Perot etalon). Fluctuations in the laser wavelength can then be held to less than 1 part in 10^8 over measurement times of a few seconds.[18,19] This procedure has the additional advantage that if the length of the reference cavity is four times the separation of the mirrors in the interferometer, phase steps of 90^0 can be introduced by locking the laser wavelength to successive resonances of the reference cavity.[20]

A problem arises from the variation of the output power of the laser as its output wavelength is shifted. As a result, the assumption normally made in the phase calculation algorithm that the intensity of the beams is constant is no longer valid. If we use a phase-calculation algorithm involving five measurements of intensity corresponding to phase shifts of -180^0, -90^0, 0, $+90^0$, and $+180^0$, respectively, and if the two beams are of equal intensity, the error in the calculated value of the phase is

$$\Delta\varphi = \varphi' - \varphi \approx (\Delta I/2) \cos \varphi \qquad (5)$$

where ΔI is the relative change in the output of the laser diode for a phase shift of 180^0. Since the phase error varies sinusoidally with the same period as the fringes, the simplest method of eliminating it is to introduce an additional phase shift of 180^0 between the beams and to record a second set of phase data.[21] The average of the two sets of phase data is then free from error. This procedure has the advantage that it also eliminates errors due to extraneous fringes and scattered light.[22]

7. CONCLUSIONS

Laser diodes are very useful light sources for interferometry because they offer low cost, compact size, low power consumption, and a stable single-mode output. In addition, the fact that the output wavelength can be controlled by varying the injection current has been utilized very effectively in several new types of interferometers. Provided care is taken to eliminate specific sources of error, high-precision measurements can be made with laser diodes.

8. References

1. S. Kobayashi, Y. Yamamoto, M. Ito and T. Kimura, "Direct Frequency Modulation in AlGaAs Semiconductor Laser," *IEEE J. Quant. Electron.* **QE-18** (4), 582-595 (1982).

2. A. Dandridge and L. Goldberg, "Current-induced frequency modulation in Diode Lasers," *Electron. Lett.* **18** (7), 302-304 (1982).

3. M. Yonemura, "Wavelength-change characteristics of semiconductor lasers and their application to holographic contouring," *Opt. Lett.* **10** (1), 1-3 (1985).

4. H. Kikuta, K. Iwata and R. Nagata, "Distance Measurement by the Wavelength Shift of Laser Diode Light," *Appl. Opt.* **25** (17), 2976-2980 (1986).

5. R. Crane, "Interference phase measurement," *Appl. Opt.* **8** (3), 538-542 (1969).

6. T. Kubota, M. Nara and T. Yoshino, "Interferometer for Measuring Displacement and Distance," *Opt. Lett.* **12** (5), 310-312 (1987).

7. T. Yoshino, M. Nara, S. Manatzakanian, B. Lee and T. Strand, "Laser Diode Feedback Interferometer for Stabilization and Diplacement Measurements," *Appl. Opt.* **26** (5), 892-897 (1987).

8. T. Suzuki, O. Sasaki and T. Maruyama, "Phase Locked Laser Diode Interferometry for Surface Profile Measurement," *Appl. Opt.* **28** (20), 4407-4410 (1989).

9. O. Sasaki and H. Okazaki, "Detection of Time-Varying Intensity Distribution with CCD Image Sensors," *Appl. Opt.* **24**, 2124-2126 (1985).

10. O. Sasaki and H. Okazaki, "Sinusoidal Phase Modulating Interferometry for Surface Profile Measurement,' *Appl. Opt.* **25** (18), 3137-3140 (1986).

11. O. Sasaki, H. Okazaki and M. Sakai, "Sinusoidal Phase Modulating Interferometry Using the Integrating Bucket Method." *Appl. Opt.* **26** (6), 1089-1093 (1987).

12. A. J. den Boef, "Interferometric Laser Rangefinder Using a Frequency Modulated Diode Laser," *Appl. Opt.* **26** (21), 4545-4550 (1987).

13. O. Sasaki, K. Takahashi and T. Suzuki, "Sinusoidal Phase Modulating Laser Diode Interferometer with Feedback Control System to Eliminate External Disturbances," *Proc. SPIE*, **1163**, 14-21 (1990).

14. J. H. Bruning, D. R. Herriott, J. E. Gallagher, D. P. Rosenfeld, A. D. White and D. J. Brangaccio, "Digital Wavefront Measuring Interferometer for Testing Optical Surfaces and Lenses," *Appl. Opt.* **13** (11), 2693-2703 (1974).

15. K. Creath, "Phase-Measurement Interferometry Techniques," in *Progress in Optics*, E. Wolf Ed., Vol. XXVI, pp. 349-393. (North-Holland, Amsterdam, 1988).

16. K. Tatsuno and Y. Tsunoda, "Diode Laser Direct Modulation Heterodyne Interferometer," *Appl. Opt.* **26** (1), 37-40 (1987).

17. Y. Ishii, J. Chen and K. Murata, "Digital Phase-Measuring Interferometry with a Tunable Laser Diode," *Opt. Lett.* **12** (4), 233-235 (1987).

18. K. Isozaki, M. Watari, E. Ogita, K. Ikezawa.and T. Ueda, "Development of a Wavelength Stabilized Optical Source with a Laser Diode," *Proc. SPIE*, **1162**, 99-106 (1989).

19. P. Hariharan, "Phase-Stepping Interferometry with Laser Diodes. 2: Effects of Laser Wavelength Modulation," *Appl. Opt.*, **28** (10), 1749-1750 1989).

20. B.K. Ward and K. Seta (in preparation).

21. P. Hariharan, "Phase-stepping Interferometry with Laser Diodes: Effect of Changes in Laser Power with Output Wavelength," *Appl. Opt.* **28** (1), 27-29 (1989).

22. J. Schwider, R. Burow, K.-E. Elssner, J. Grzanna, R. Spolaczyk and K. Merkel, "Digital Wave-Front Measuring Interferometry: Some Systematic Error Sources," *Appl. Opt.* **22** (21), 3421-3432 (1983).

New methods for economic production of prisms and lenses

G. Richter

Wilhelm Loh Wetzlar, Optikmaschinen GmbH & Co. KG
D - 6330 Wetzlar / Germany

1. Prism manufacturing in small batches at reduced cost

1.1 General

The dramatic increase in costs for rough grinding or milling of prisms is due to the production of smaller batches and the variety in the number of types to be produced.

We were thus faced with the task of searching for improved economical methods which would assure the user flexibility and a competitive position on the market.

LOH, thereafter, developed an advanced rough milling technology for prisms which consists of computer-controlled data, a newly designed milling machine and the appropriate CNC control unit.

Many factors contribute to the substantial savings in production costs. The examples (Fig. 1) indicate a saving of 76 % and 69 %.

The savings result from:

- Simplified Clamping Jigs
- Less Set-Up Time
- Shorter Processing Time
- Elimination of Procedures (e.g. cementing on to jigs)
- Fewer Tools
- Computer Support
- Easy Training of Personnel
- Easier Repeatibility
- Increased Number of Work Stages in One Working Process
- Greater Flexibility

Some of these points are explained in greater detail in the following sections.

1.2 Computer Preparation (Fig. 2)

The component dimensions are to be entered in a register, transfered to a disk over a PC or entered directly into the CNC Control. If the same workpieces are to be produced again later on, transfer the data from the disk and the actual tool wear only may be corrected.

1.3 **Clamping Jigs** (Fig. 3)

Workpiece fixtures are extremely simple and can be easily accomplished by the user. The change-over takes a maximum of 2 minutes to complete. In special cases, the use of inserting device is recommended at low oversize.

1.4 **Grinding Wheels**

Only two tools are required for the operation of the LOH Prism Milling Machine CNC: one with small lips for plano surfaces and one with wide lips for workpieces with plano surfaces and roundings.

1.5 **Set-Up-Times**

The set-up-times are very short and can be conducted in less than 10 minutes. The chucking of the workpiece is always done mechanically. Due to this fact, warming up, cementing, decementing and cleaning; as well as the costs for additional jigs, are eliminated.

1.6 **Prism Milling Machine CNC** (Fig. 4)

The Prism Milling Machine CNC, in accordance to its' basic construction, is a 3-axis machine. These axis are integrated into a torsion-rigid casting frame.

The work table (X-axis) is moved horizontically and is led through rollers. The drive results over a gear-controlled stepping motor. This motor brings forth the feed during plano-operations and controls the contact edge to the reference diameter of the tool during roundings.

The Z-axis is situated in a 90° angle to the X-axis. The carriage contains roller guideways and the adjustment is untertaken through a stepping motor over a planetary roll-spindle with an incline of 2 mm. The tool is brought into position before the plano-operation begins and is maintained at this position. The tool is constantly controlled in accordance to the rounding measurements during rounding.

A vertically-installed turn table (Y-axis) is found in the work table. The drive results over a gear-controlled stepping motor. An easily accomplished workpiece fixture, which is matched to each workpiece, is secured to the turn table. The specific swivel-angles of the prisms to be produced are undertaken through this process. The exactness lies at 1 angular minute. This position is held during the process with an electromagnetic brake. The turn table is constantly controlled during roundings through the interaction of the X-axis and the Z-axis. An incremental rotary encoder has been built in for position control.

The workpiece is held over a clamping bolt. Operations results hydro-pneumatically over a foot wsitch. The height of the chucking cylinder is adjustable and hydraulically clamped.

The working area of the machine is covered with a splash-guard. This guard is opened and closed automatically during automatic operations.

The processing of all surfaces positioned in one level results in <u>one</u> clamping. Plano surfaces, as well as plano surfaces combined with roundings, can be processed with the Prism Milling Machine CNC. Up to 64 work steps can take place one after another. The largest surface should not exceed 100 mm.

1.7 CNC control unit

The CNC control of the Prism Milling Machine is specially developed and works in dialog with the operator. Work shop programming and well thought out monitor construction allows simple operation and fast training of personnel. A program language is not required.

Operation is subdivided under "Machine setting", "Manual Operation" and "Automatic Operation".

The monitor first offers an input menu consisting of

- Machine and Tool Data
- Prism Data
- Roundings
- Disks - Input - Output
- Automatic

The corresponding input data will appear when the operator requests machine and tool data.

I N P U T: Machine / Tool	**L O H W E T Z L A R**	**Prism Milling Machine CNC**
Tool Data:	Outside Diameter...............DA[mm]	100
	Inside Diameter ....DI[mm]	70
	Tool Reference Diameter........DB[mm]	75
Machine Data:	Reference Dimension............RM[mm]	80
	Tool Correction (Magnitude)....WK[mm]	0.2
	Rotation Axis Reference Correction.................[degrees]	0.17
Speed:	Operating Feed..............[mm/min]	100
	Fast Drive.........................[%]	100

The cursor can be placed to the desired position with the cursor keys. All values are entered into the memory by using the PA-key.

The monitor with workpiece data is identical to the chart indicated in Fig. 2.

The machine calibration during automatic operation takes place before operation of the first workpiece begins.

Each operation with all associated parameters appears on the monitor and are easily controlled by the user.

A U T O M A T I C: Operation — **L O H W E T Z L A R** — **Prism Milling Machine CNC**

Plano surfaces

WI..[grd]	H..[mm]	E..[mm]	A..[mm]	B..[mm]	V..[mm/min]
90	28	19,50	42	32	100

Roundings

Amount:	R..[mm]	M1..[mm]	M2..[mm]
	19,501	19,501	19,50

1.8 **Operating Examples**

Some typical operating examples are shown in Fig. 5. The layouts show the 1st and 2nd work steps respectively for 3 prisms.

1.9 **Summary**

The new Prism Milling Method with CNC controlled machine offers greater advantages in comparison with present processing methods for smaller batches through:

- lower costs for tools and clamping jigs,
- fewer work steps,
- computer-supported processing,
- greater flexibility through extremely short set-up-times,
- easier repetition through data storage on disks,
- easy operation for newly instructed personnel.

2. New developments for economic centering of high precision optics

2.1 General

Present day centering machines are not suitable for the constantly increasing demands accuracy necessary for the future. Above all, the production of electronic components and lithography require optical systems with highly accurate images. Therefore, extremely precise lenses are required.

Due to this fact, LOH developed a new machine concept "CENTROMATIC CNC" for centering lenses in the visual and infrared ranges of application. These new developments are based on the experience obtained through the some 2000 machines that have been built thus far and in association with the latest, state-of-the-Art technological discoveries.

Our goal was:

- to meet the highest qualitiy demands
- suitability for large and small series
- drastic reduction in set-up and processing times
- lower costs for tools
- economical processing of complicated lens shapes

The advantage over presently used machines are:

2.2 Construction (Fig. 6)

Straight-lined movements can be obtained through the elimination of all controlling cams. The honeycombed cast design guarantees high rigidity and the built-in vibration absorbers prevent the influence of external vibrations.

The patented drive of the centering spindles results through the use of flat belts (no more gear wheels). The synchronous drive is obtained through the use of a sensitive differential. The result is a reduction in centering chuck marks.

2.3 Control axis

The 6-axis control consists of:

The two X - Y axis of the grinding spindles, which allow individual adjustment of the fast and normal processing movements. The center thickness adjustment compensates for the various lens thicknesses and guarantees equal bevel sizes, by request. The electronic diameter measuring system has a resolution of 1 micron and allows corrections for tool wear and temperature changes.

2.4 Centering spindles

The upper and lower centering spindles are identical and have a diameter of 60 mm and high rigidity. The runout lies within 0,001 mm. Easy exchange of the spindles is guaranteed.

2.5 Hydrostatic clamping (Fig. 7)

The lower centering spindle is sensitively slided in simultaneous vibration in the direction of the axis. This allows easier internal alignment of the lenses with clamping angle in some cases also under 10°. The clamping results hydrostatically without axis misalignment and wear.

2.6 Fitting of centering chucks and grinding tools (Fig. 8)

The diamond tools are mounted in a quick chucking device which reduces the tool change time up to 50 % and obtains concentricity within 0,01 mm.

The same system is used for the fitting of mandrels (no more screwing!) and guarantees repetition by changes in rounding and plano runout within 0,001 mm.

Considerable savings in set-up-time and especially with smaller lot sizes are made possible through these features.

2.7 New grinding tool geometry (Fig. 9 + 10)

For the first time ever, a new grinding tool geometry allows the direct dressing of all standard tools without dismantling the cylinder and bevel components. Also the new geometry results in a substantial reduction of tools necessary for the different lens shapes.

2.8 Laser centering (Fig. 11)

Lenses which cannot be centered with the bell clamping process are centered with the LOH Laser Centering Device. This allows the economic centering of such lenses without cementing.

The monitor displays a read-out and graph of the surface angle tilt according to DIN 3140 on the base of the input of refraction index of the lens, the pre-selected amplification and the deviation of the laser beam measures.

The Laser Centering Device can be operated in transmitted light and reflection situations. The capsuled-construction of the focusing optic protects the system from oil mist and other sorts of pollution.

2.9 Loading

The LOH Centering Machine 240 CNC has a defined interface which can be applied to manual, as well as automatic loading procedures.

2.10 CNC-Control "LOHTRONIC 100"

The control of the Centromatic 240 CNC has more obvious advantages over standard type controls:

- Simple programming in clear text.
- Dialog-controlled parameter programming through graphic designs of tools for reduction in programming times.
- Dynamic process simulation on the graphic monitor for input inspection.
- Optimal machine-emulation program in support of process preparation and the speeding up of set-up-times.
- Laser-controlled alignment of the lenses with numerical indication of the surface angle tilt.
- Machine-integrated measuring stations for ascertainment of self-processed corrective data at two levels, with programmable measuring cycles.
- Internal logic for the plausibility examination and prevention of input mistakes.
- The data obtained during teach-in-operation can be directly programmed into the computer.
- Set-up data can be stored on a disk; a measure that drastically reduces set-up-times.
- Practice-orientated monitor graphics for adjustment, simulation and processing steps.

2.11 Advantages of the Centromatic 240 CNC

The following is another short summary of advantages:

- Shorter set-up and processing times (Fig. 12)
- Optimized grinding process (no cams)
- Flat grinding (higher RPM) and deep milling (lower RPM) possible
- Stronger and identical centering spindles
- Automatic Splash-Guard
- Fewer specially-contoured wheels
- New tool geometry (no dismantling during dressing)
- Better tool clamping with higher accuracy
- Centering chuck and special clamping sleeve attached (no threads, meaning no runout by chuck changes)
- Hydrostatic spindle clamping
- Automatic centering at smaller clamping angles
- Direct indication of the surface angle tilt
- Highest diameter-accuracy through automatic compensation of tool wear

LOH

COST COMPARISON

PRISM CNC

(83.72)
(0.27)
2.75
28-0.2
45°
135°±2'
39.4-0.2
19.5
39-0.05
8×45°
80.2-0.2

PRISM 1

90°
120°±3'
26±0.2
(45.03)
26-0.2
26
90±2'

PRISM 2

%
100
80
60
40
20
Prism 1
Prism 2
24
31
1
2
3

COSTS FOR MILLING

PRODUCTION - METHODS

1 PRESENTLY

2 + 3 PRISM MILLING CNC-MACHINE

FIG. 1

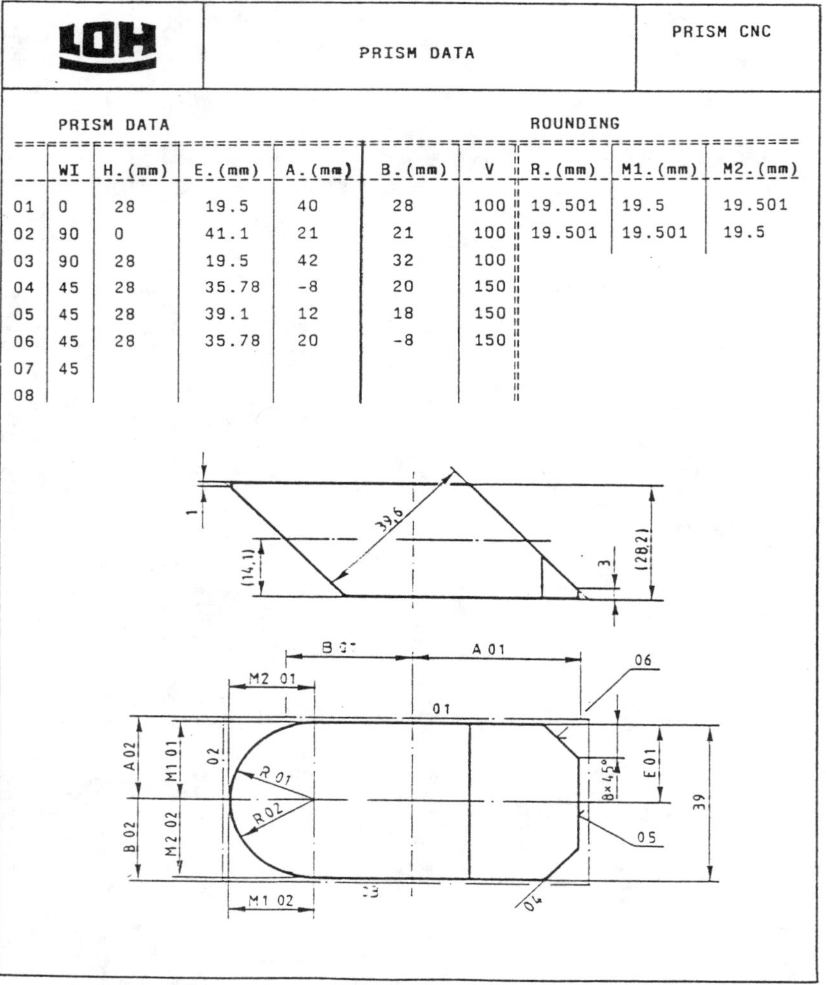

PRISM DATA / ROUNDING

	WI	H. (mm)	E. (mm)	A. (mm)	B. (mm)	V	R. (mm)	M1. (mm)	M2. (mm)
01	0	28	19.5	40	28	100	19.501	19.5	19.501
02	90	0	41.1	21	21	100	19.501	19.501	19.5
03	90	28	19.5	42	32	100			
04	45	28	35.78	-8	20	150			
05	45	28	39.1	12	18	150			
06	45	28	35.78	20	-8	150			
07	45								
08									

FIG. 2

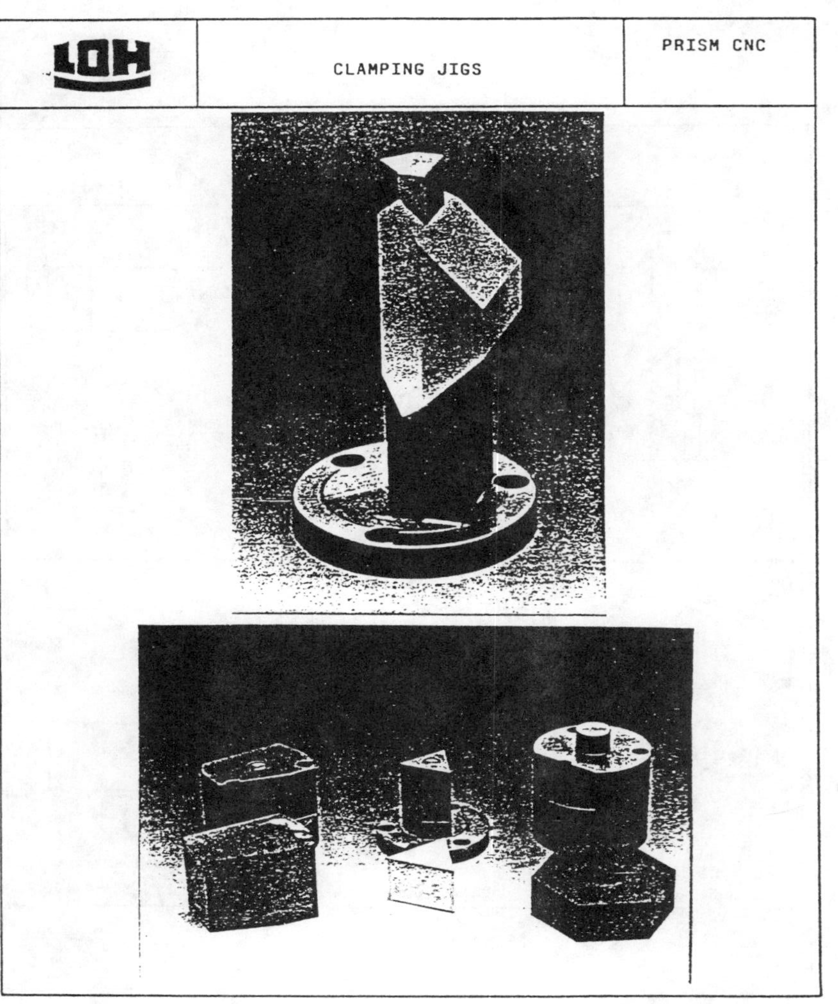
LOH
CLAMPING JIGS
PRISM CNC

FIG. 3

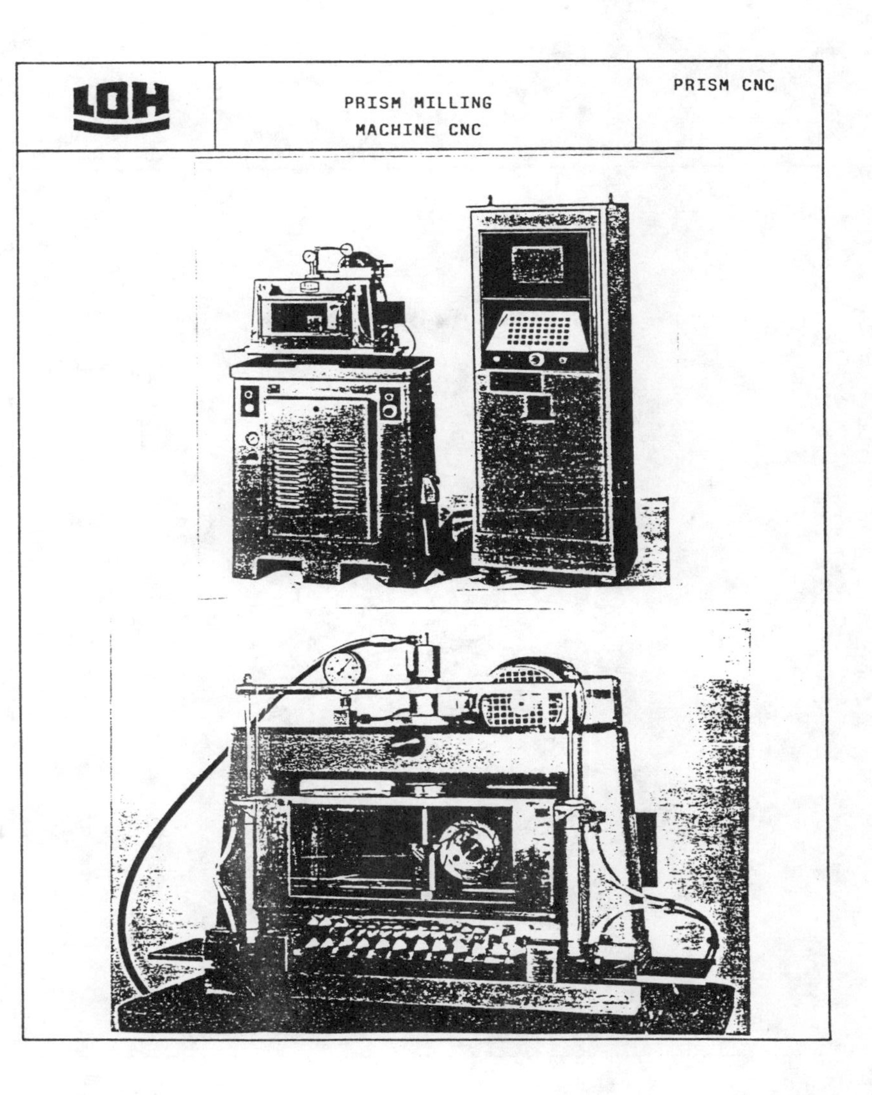
LOH
PRISM MILLING
MACHINE CNC
PRISM CNC

FIG. 4

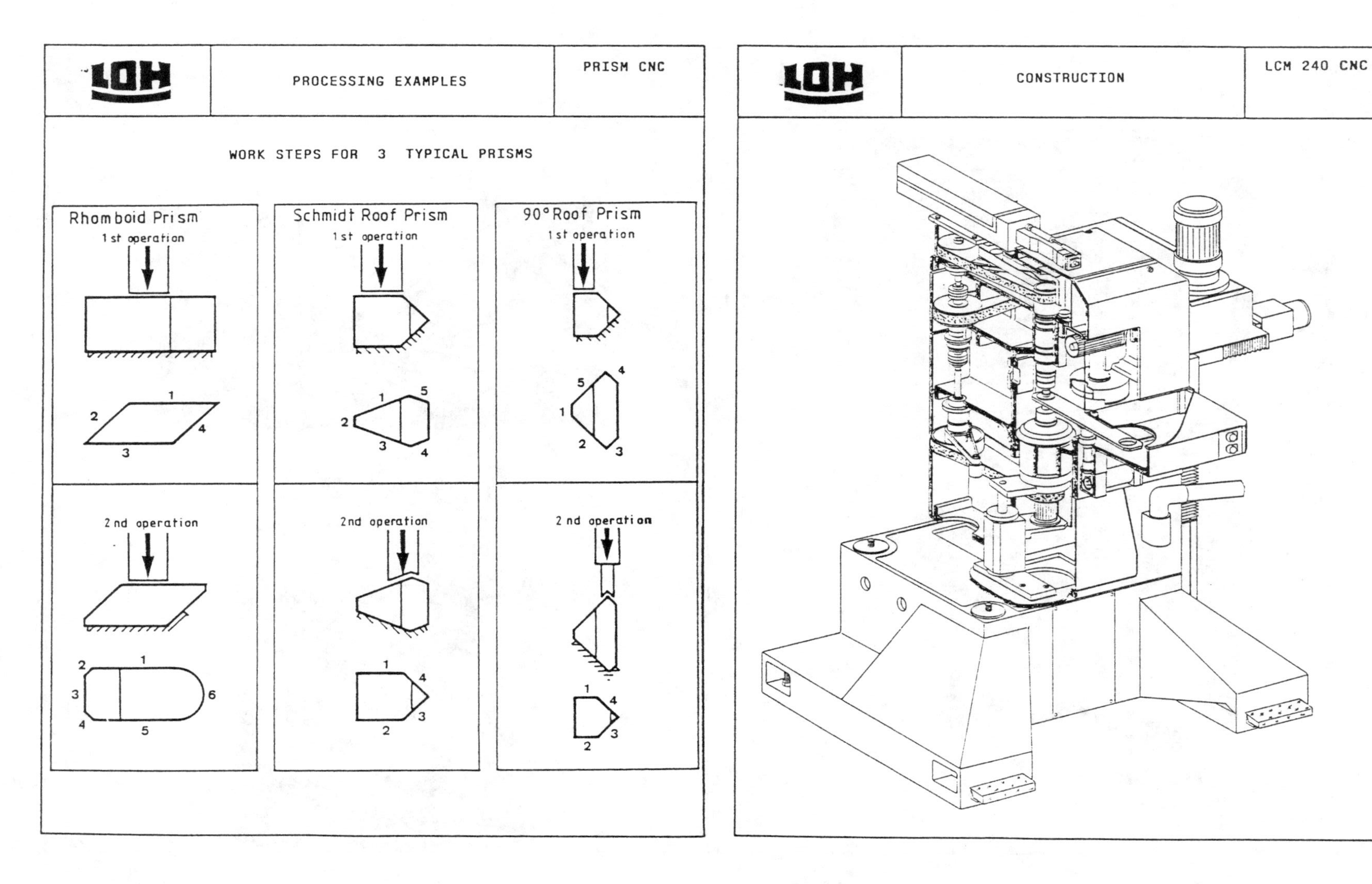
LOH
PROCESSING EXAMPLES
PRISM CNC
WORK STEPS FOR 3 TYPICAL PRISMS
Rhomboid Prism
1st operation
2nd operation
Schmidt Roof Prism
1st operation
2nd operation
90° Roof Prism
1st operation
2nd operation
LOH
CONSTRUCTION
LCM 240 CNC

FIG. 5

FIG. 6

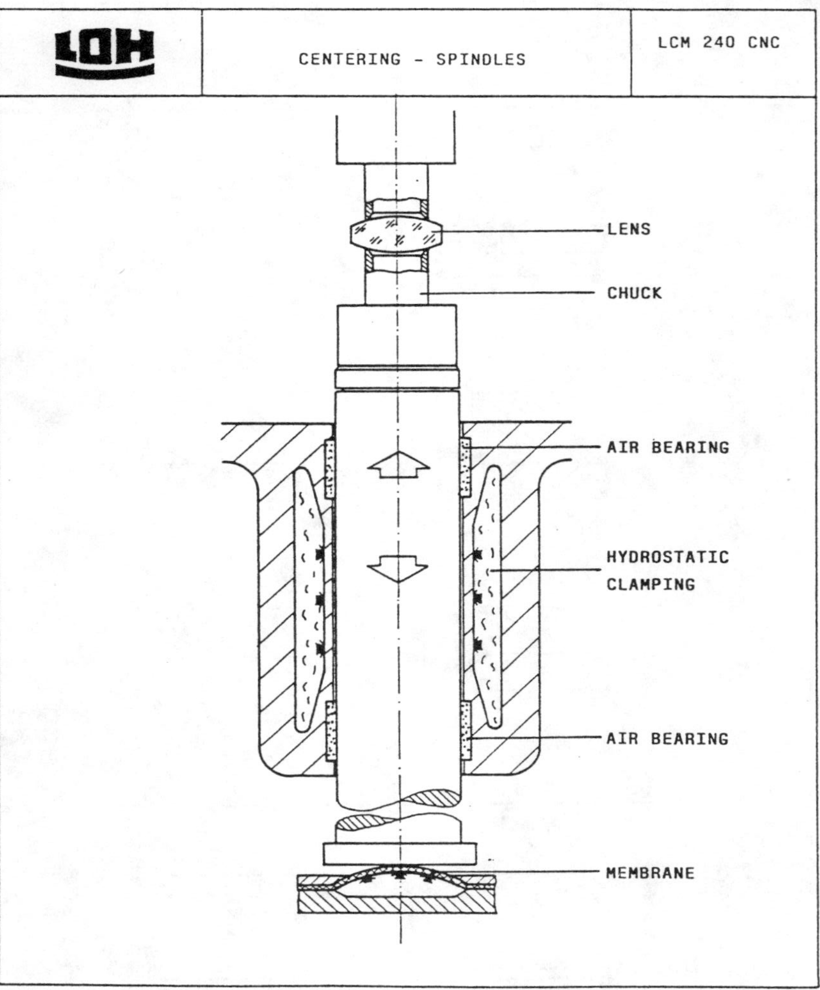

LOH
CENTERING - SPINDLES
LCM 240 CNC
LENS
CHUCK
AIR BEARING
HYDROSTATIC
CLAMPING
AIR BEARING
MEMBRANE

FIG. 7

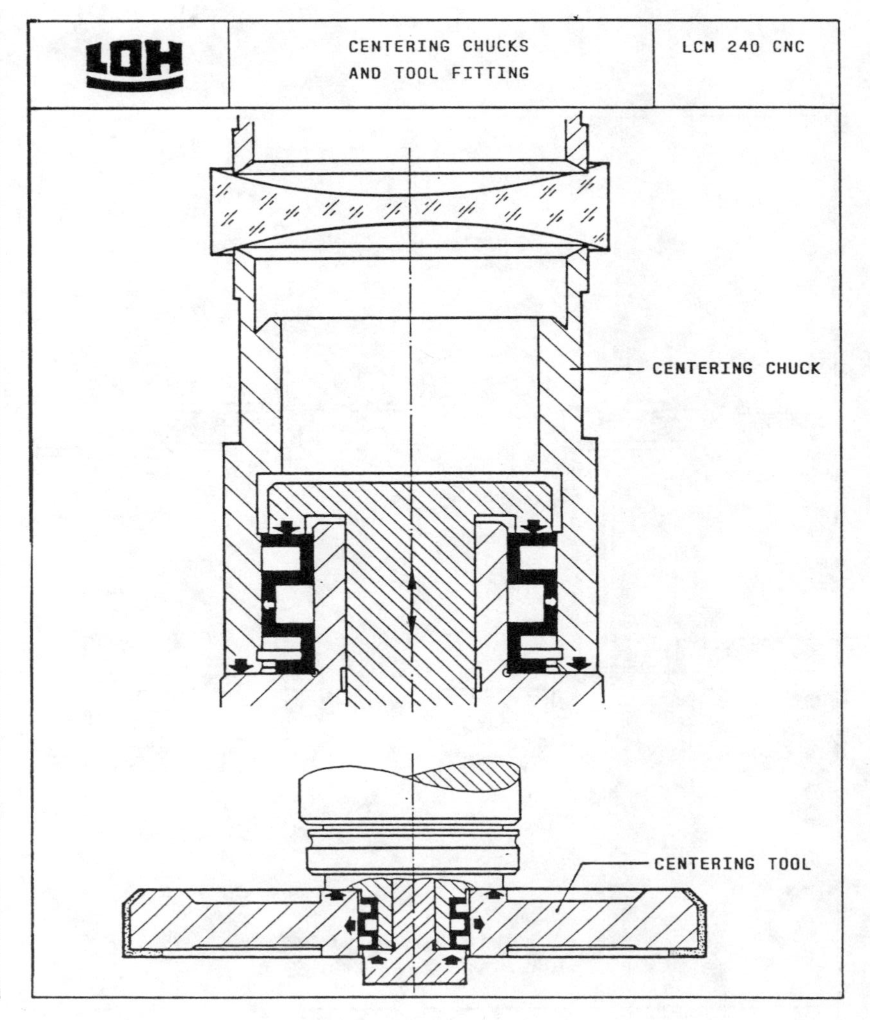

LOH
CENTERING CHUCKS
AND TOOL FITTING
LCM 240 CNC
CENTERING CHUCK
CENTERING TOOL

FIG. 8

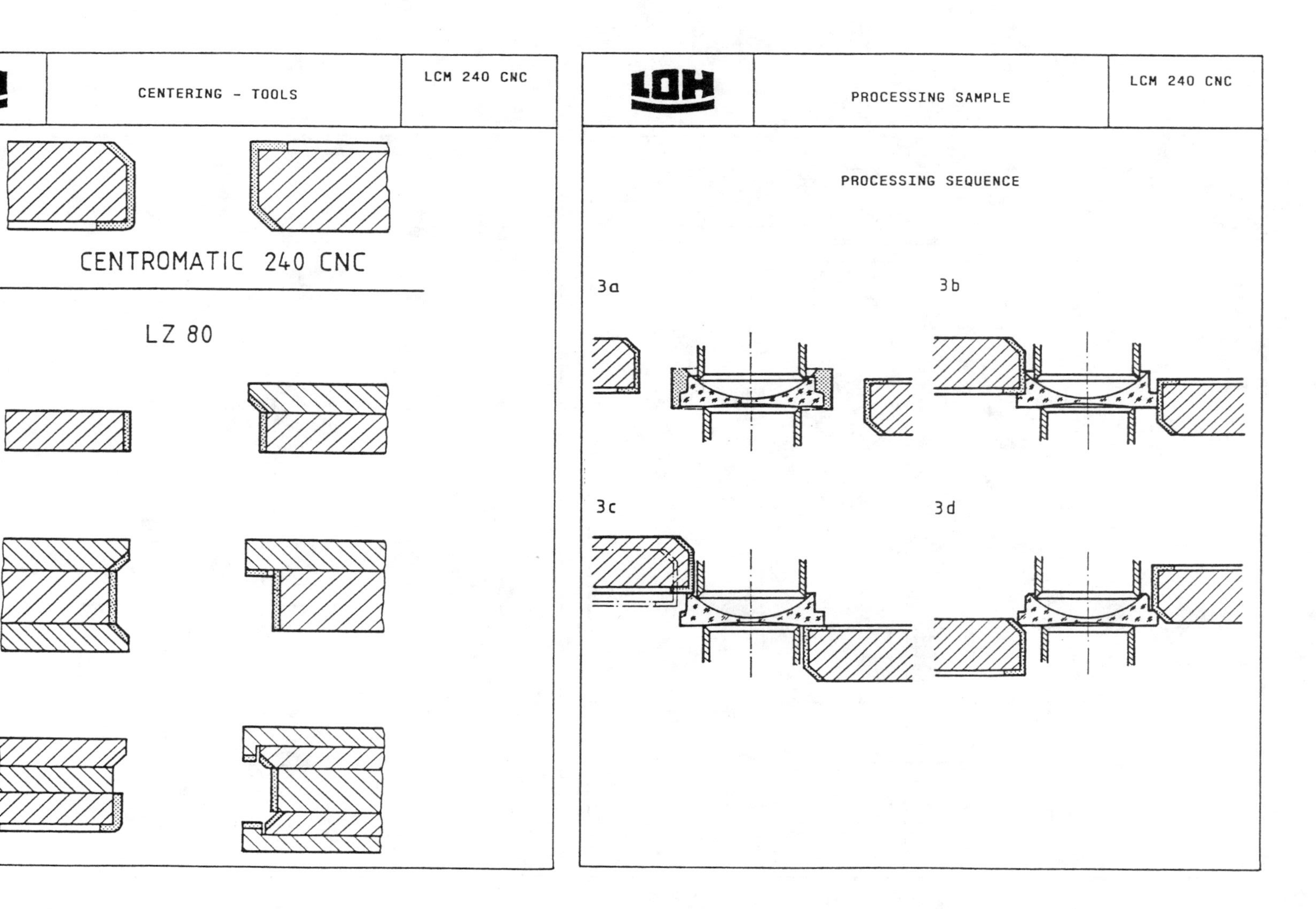
LOH
CENTERING - TOOLS
LCM 240 CNC
CENTROMATIC 240 CNC
LZ 80
LOH
PROCESSING SAMPLE
LCM 240 CNC
PROCESSING SEQUENCE
3a
3b
3c
3d

FIG. 9

FIG. 10

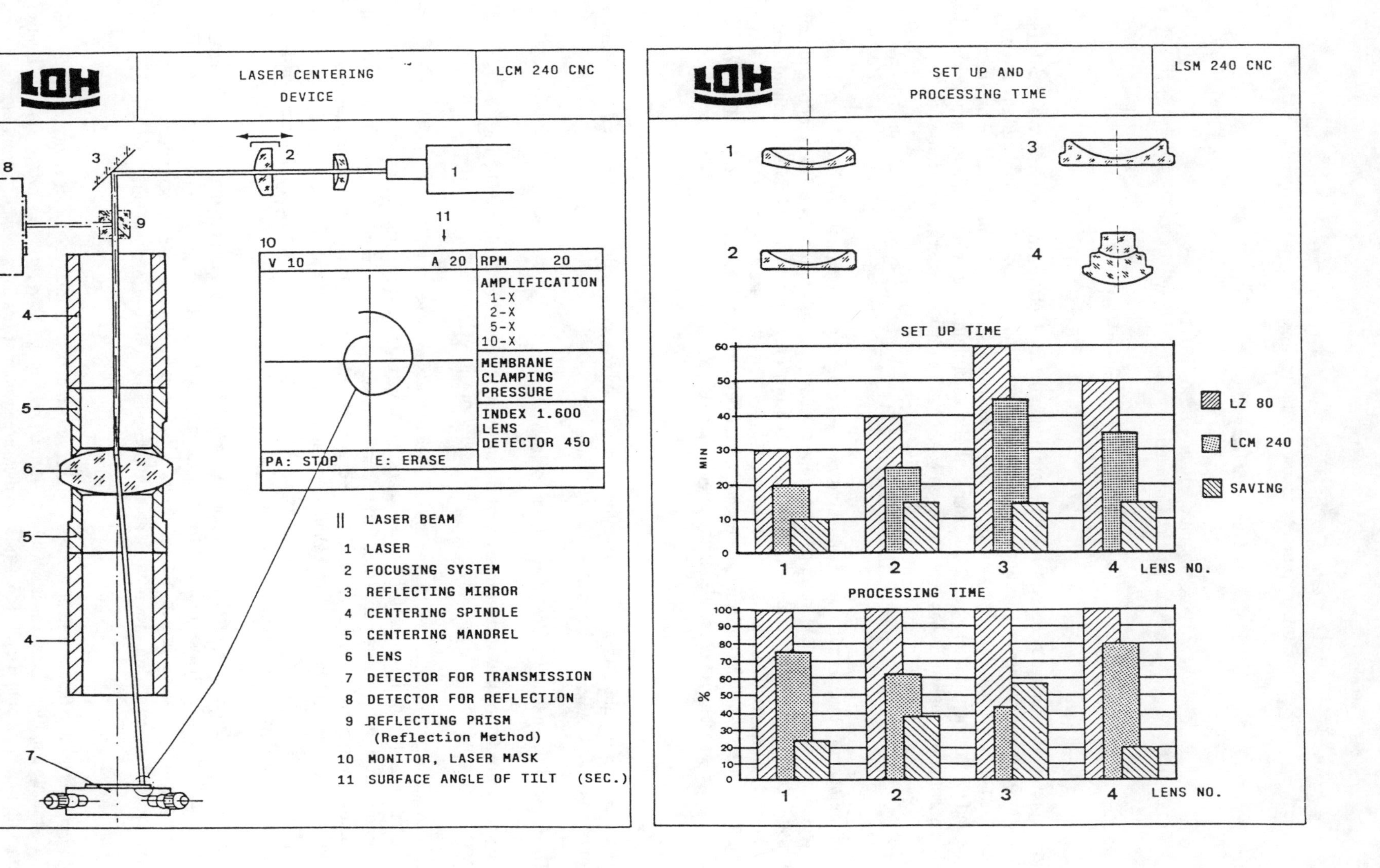
LOH
LASER CENTERING DEVICE
LCM 240 CNC
V 10
A 20
RPM 20
AMPLIFICATION
1-X
2-X
5-X
10-X
MEMBRANE CLAMPING PRESSURE
INDEX 1.600
LENS
DETECTOR 450
PA: STOP
E: ERASE
LASER BEAM
1 LASER
2 FOCUSING SYSTEM
3 REFLECTING MIRROR
4 CENTERING SPINDLE
5 CENTERING MANDREL
6 LENS
7 DETECTOR FOR TRANSMISSION
8 DETECTOR FOR REFLECTION
9 REFLECTING PRISM (Reflection Method)
10 MONITOR, LASER MASK
11 SURFACE ANGLE OF TILT (SEC.)
LOH
SET UP AND PROCESSING TIME
LSM 240 CNC
SET UP TIME
MIN
LENS NO.
LZ 80
LCM 240
SAVING
PROCESSING TIME
%
LENS NO.

FIG. 11

FIG. 12

Interferometer accuracy and precision

Lars A. Selberg

Zygo Corporation, Laurel Brook Road
Middlefield, Connecticut 06455

ABSTRACT

Several optical errors present in interferometer systems are examined in detail. Optical cavity errors are typically the primary limitation on measurement accuracy. Secondary sources of error include imaging distortion and ray-mapping (slope) errors. The effect of phase modulation/processing errors on precision is briefly discussed. Some on-site errors (operator technique and environmental effects) are also examined. Rules of thumb for error estimation are presented when applicable. Specific quantitative analysis is based on the Zygo MARK IV*xp* phase measuring interferometer.

1. INTRODUCTION

Interferometric measurement of optical surfaces and optical assemblies can achieve accuracies of λ/50 PV or better if careful metrology technique is observed. It is incumbent upon the optician and QC engineer, i.e., the vendor and the customer, to understand the sources of error and thus the limits of their instrumentation vs. the limits of technique and environment. The optical factors contributing to errors in an interferometer system are analyzed in as generic a treatment as possible. Quantitative analysis which is instrument dependent has been directed toward the Zygo MARK IV*xp* interferometer. Error sources specific to phase measuring interferometry are summarized.

1.1 Terminology[1,2]

Quaesitum - The true value of the quantity which is being measured. The error in the optic under test *is* the quaesitum, not part of the error in our measurement.

Measurement - The value of the quantity as determined from some instrument.

Accuracy - The degree to which the measurement represents the quaesitum. Accuracy includes all error sources. Comparisons of accuracy should be made consistently, i.e., PV vs. PV or rms vs. rms.

Precision (repeatability, dubiety) - Precision represents the variance of measurement. It includes all random and cyclic errors. It is a statistical quantity which is not related to the quaesitum. Instrumental precision indicates the ability of the interferometer to produce the same measurement value for the same measurement conditions. Experimental precision allows for changes in the measurement conditions as might be expected under "normal" use. It is important to distinguish between the precision of measurement (actual phase calculation), and the precision of quantities derived from the measurement (e.g., Peak-Valley, RMS, 3rd order aberrations, etc.). As with accuracy comparisons of precision should be made consistently, i.e., PV vs. PV or rms vs. rms.

Resolution - As applied to interferometers, resolution might refer to the number of bits used to digitize interferogram intensities or final phase values, or spatial sampling of the detector array for optical resolution. For example, the MARK IV*xp* digitizes phase to 15 bits per fringe, or a resolution of λ/65536 if used for double pass measurements. This implies that the digitization error is very small, but says nothing about magnitudes of other errors (typically in the range of λ/100 to λ/20).

Reference - Most often it is used to denote one of the two surfaces which define an interferometer cavity, the *reference surface*. It can also indicate a mathematical surface which is fit to a phase map, the *reference plane or sphere*. Lastly, it is used in connection with measurements which are intended to measure system errors for subtraction from measurements of test pieces, *reference measurement and subtraction*.

2. CAVITY ERRORS

An interferometer measures an interferometric cavity, not a single surface or lens. The simplest interferometer cavity, C, is defined by the test, T, and reference, R, surfaces,

$$C = T \pm R \tag{1}$$

where the sign is defined by the type of interferometer[3]. This is a relative measurement, i.e., the test surface is being measured relative to the reference surface. The surface deviations in the test surface comprise the quaesitum, while the surface deviations of the reference surface are a source of error known as cavity errors. To first order, the accuracy of interferometric measurement is limited by the cavity errors. Absolute testing[3-8] and reference subtraction[5] techniques can be used to eliminate, or at least reduce, cavity errors. In the simple two surface cavity, an interferogram represents either the sum (e.g., Twyman-Green) or difference (e.g., Fizeau) of the surfaces (see Figure 1 a and b).

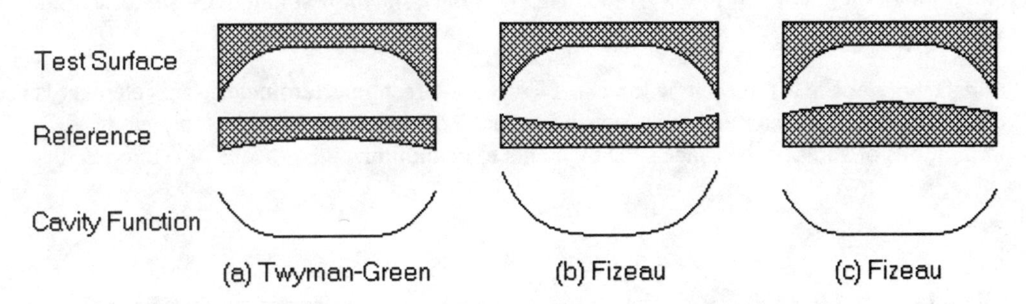

FIGURE 1. PLANO INTERFEROMETER CAVITIES SHOWING A $\lambda/4$ PV PART UNDER TEST (TOP), $\lambda/20$ PV REFERENCE SURFACE (MIDDLE) AND CAVITY FUNCTION (BOTTOM). THE PV STATISTICS OF THE CAVITIES ARE (A) 0.2λ, (B)0.3λ PV AND (C) 0.2λ PV.

This illustrates that the same reference used in two types of interferometer can yield significantly different results, making the test surface look better or worse than it actually is. The same ambiguity is seen for a single type of interferometer when the reference surface is specified by a single value, e.g., $\lambda/20$ PV, which states nothing about the surface shape. Figure 1 b and c show an example of two reference surfaces of the same certified quality, same surface function (power), but opposite polarity. The $\lambda/5$ PV test surface is measured as 0.3λ and 0.2λ in the two cases, a discrepancy of $\lambda/10$. In general we can write the worst-case error margin for relative measurement[9] as Quaesitum = Measurement ± Error,

$$PV_T = PV_C \pm PV_R \tag{2a}$$

$$rms_T = rms_C \pm rms_R \tag{2b}$$

where the subscript T, C and R refer to the test surface, the cavity function and the reference surface, respectively. This analysis of cavity errors is equally applicable to plano and spherical testing. From this one may conclude that with relative testing using commercially available reference flats and spheres ($\lambda/20$ - $\lambda/40$ PV), $\lambda/10$ - $\lambda/20$ PV is the practical limitation on accuracy. Note that the accuracy limits of the rms statistic (2b) are similar to the PV statistic (2a). This is because the reference and test surface shapes will correlate to some degree.

For more complex cavities, equations 2 can be generalized to include all elements and surfaces which are required to construct the cavity, but are not part of the part/system under test, in the "reference" error. It may be necessary to examine carefully the scale factor used for converting fringes to waves when calculating the contribution of each surface. An example of additional surfaces is the return flat/sphere required to test samples in transmission (see Figure 2a).

Another example is spherical testing with a Fizeau interferometer (see Figure 2b). In one case (a) a plano reference and diverger lens, the wavefront aberrations of the diverger (surface errors, element tilts and decenters) and the plano reference surface errors comprise the cavity error of the measurement. In the other case (b) a diverger lens with a spherical reference as the last surface, the cavity error consists only of the surface errors of the reference surface. Case (a) is effectively identical to spherical testing with a Twyman-Green interferometer.

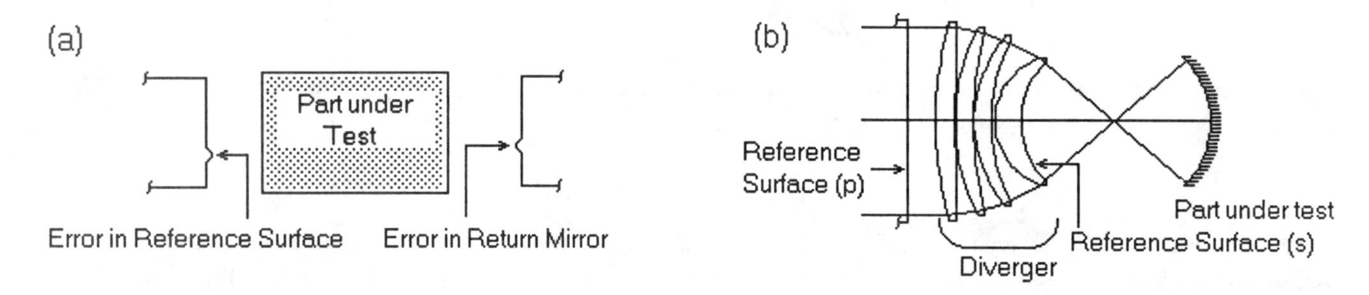

FIGURE 2. ADDITIONAL SURFACES CONTRIBUTE CAVITY ERRORS. WINDOW TESTED IN TRANSMISSION (A) REQUIRES A RETURN MIRROR (SHOWN AS FIZEAU CAVITY). SURFACE ERRORS IN THE RETURN MIRROR CONTRIBUTE TO THE CAVITY ERROR. A DIVERGER LENS ADDS TO CAVITY ERRORS (B). (P) THE DIVERGER IS WHOLLY IN THE CAVITY, CAUSING ITS TRANSMITTED WAVEFRONT ERRORS TO CONTRIBUTE DIRECTLY TO THE CAVITY ERROR. (S) ONLY THE LAST SURFACE OF THE DIVERGER CONTRIBUTES TO THE CAVITY ERROR.

A third example of additional cavity errors is the beam splitter in a Twyman-Green interferometer. This element is used differently for the two arms of the interferometer and thus contributes to the cavity error. Some testing geometries errors allow the cavity error contributions of additional surfaces and elements to be minimized by means of reference measurement and subtraction.

2.1 Customer vs. vendor

Equations 2 imply that if optics are qualified by relative testing in a customer/vendor relationship, that the specifications to which each party inspects the parts must be different. For example, a vendor is fabricating 0.2λ PV flats using a 0.05λ PV reference. The customer inspects these parts, also using a 0.05λ PV reference. In order to insure acceptance of good parts at both locations, the vendor must ship only parts which are better than 0.15λ and the customer must accept anything which is better than 0.25λ. Of course, the customer and vendor could also arrange to purchase a matched set of references, thereby reducing this difference in specification.

2.2 A priori knowledge

It is sometimes possible to improve on equations 2 by using *a priori* knowledge of the surface functions and the aperture used. For example, if the reference is qualified as $\lambda/20$ PV, but is known to be dominated by roll-off at the edge of the aperture, then the quality of the reference may be significantly better when testing smaller parts. Knowledge of the concave/convex ambiguity of the reference surface can be used to limit the error estimate to one side or the other. For example, if both surfaces are known to be concave (convex), then with a Fizeau interferometer the test surface is most likely better than the cavity function, $PV_C - PV_R \leq PV_T \leq PV_C$. This assumes that the surfaces are dominated by power or higher order roll-off and not by astigmatism. In the case of the rms statistic, if the two surface functions, R and T, are uncorrelated, e.g., a diamond turned flat with no net power or roll-off tested against a polished flat, the rms statistics of the surfaces will root-sum-square (RSS) to give the cavity rms. This may be inverted to yield,

$$rms_T \approx \sqrt{rms_C^2 - rms_R^2} \tag{3}$$

The error limits are still valid, but the use of a priori information indicates where the rms can be expected to lie within those limits. For example, if a cavity is measured as having an rms of 0.024λ using a reference with an rms of 0.010λ then $rms_T = 0.024 \pm 0.010\lambda$. If the two surface functions are uncorrelated, then $rms_T \approx 0.022\lambda$.

2.3 Reducing/eliminating cavity errors

Reference subtraction and absolute testing combine more than one relative measurement in order to reduce cavity errors. For example, the cavity in figure 2 can be measured without the window as a reference and then subtracted from the cavity as measured including the window. Such a procedure is limited only by the experimental precision of the system.

Absolute testing combines several relative measurements to eliminate the cavity errors. The three flat test[3,4] yields absolute an surface profile along a diameter of the test optic. Absolute spherical measurement[5,6] yields a map of the entire surface. Part rotation and Zernike analysis[7,8] has also been demonstrated as a technique for eliminating cavity errors, but is limited to measuring non-rotationally symmetric aberrations. These techniques eliminate cavity errors from the measurement, but can increase some of the secondary error sources because of the multiple measurements.

Most useful is the absolute measurements of references (transmission flats/spheres) which are stored for subtraction from future measurements. Spatial registration of the reference phase maps to the new measurements is very important.

3. IMAGING DISTORTION

An interferometer generally suffers some imaging distortion, especially when using fast diverger lenses for testing spheres. This is dominated by the sinθ mapping of most lenses, as opposed to $k\theta$ imaging (ideal for spherical testing). Distortion stretches the phase map of the part under test unevenly. This distortion is approximately given by,

$$\gamma \approx \frac{\sin\alpha_{max}}{\alpha_{max}} - 1 \propto \alpha_{max}^{2} \tag{4}$$

where α_{max} is the half angle of the transmission sphere (diverger lens), i.e., 1 / (2 f/#). For plano interferometers the imaging system will also have some small amount of distortion (for the MARK IV*xp* this is < 1% at 102 mm aperture). Distortion stretches planes and spheres (tilt and focus) into other (aberrated) shapes. The error due to distortion is proportional to the phase slope, i.e., fringe density, indicating *that highest accuracy testing of optics is done with the cavity nulled.* For systems which require a large number of fringes, such as fringe measuring systems, distortion can be a significant problem and calibration for distortion is strongly advised. For some testing, such as aspheres, it is impossible to null the fringe pattern without a null lens or compensator plate. The distortion will then cause an error in the shape of the cavity function. Consider a perfect cavity with some amount of tilt and focus error,

$$\mathrm{OPD}\,(r,\theta) = a\,r\cos\theta + c\,r^{2} \tag{5}$$

where r and θ are the normalized coordinates on the test piece, and a and c are the tilt and power coefficients; for simplicity, tilt is taken as being in the x-direction only. Distortion maps the radial coordinate in a non-linear fashion. The distortion of equation 4 is approximated as third order distortion, $r \rightarrow r\,(1 + \gamma r^{2})$, where γ is the third order distortion at the full test aperture (r=1), defined as the fractional real ray deviation from the corresponding paraxial ray[10]. The distorted phase map is,

$$\mathrm{OPD}\,(r,\theta) = a\,r\cos\theta + c\,r^{2} + a\,\gamma\,r^{3}\cos\theta + c\,\gamma\,r^{4} + c\,\gamma^{2}\,r^{6} \tag{6}$$

The distortion has created terms which appear as third order coma, third order spherical and fifth order spherical. Fitting a reference sphere to this distorted phase map and analyzing for PV and RMS errors, one can derive the following rules of thumb,

	tilt only	power only
$PV_{error} \approx$	$\dfrac{a\,\gamma\,r_T^{2}}{6}$	$\dfrac{c\,\gamma\,r_T^{2}}{4}$
$rms_{error} \approx$	$\dfrac{a\,\gamma\,r_T^{2}}{14}$	$\dfrac{c\,\gamma\,r_T^{2}}{11}$

(7)

where a is the tilt in fringes over the part *diameter*, c is the power in fringes over the part *radius*, PV_{error} and rms_{error} are in waves, assuming a double-pass measurement, and r_T is the radius of the part under test as a fraction of the radius for which γ was defined. For combinations of tilt and power, the expressions in equation 7 can be root-sum-squared for PV or rms errors.

Distortion errors are of concern primarily for fast spherical testing. These errors are directly proportional to the fringe density of the interferogram and are thus eliminated at null. If testing mild aspheres without compensator plates or null lenses, the measured shape of the surface will be affected by this distortion. Any part which is smaller than the interferometer aperture should be centered in the aperture to minimize distortion effects.

	Tilt only (10 fringes)		Focus only (2 rings)	
Cavity	PV	rms	PV	rms
plano	0.006	0.0022	0.002	0.0009
f/7.2	0.008	0.0025	0.003	0.0010
f/0.75	0.131	0.0455	0.042	0.0179

TABLE 1. DISTORTION ERRORS FOR THE ZYGO MARK IV*XP* 102 MM PLANO, F/7.2 SPHERICAL AND F/.75 SPHERICAL CAVITIES USED AT FULL APERTURE. THIS EXAMPLE ASSUMES 10 TILT FRINGES OVER THE DIAMETER (LEFT) AND 2 RINGS FOCUS ERROR (RIGHT).

4. RAY-MAPPING ERRORS

Ray-mapping errors are caused by wavefront slopes in the interferometer source optics, surface slope errors in surfaces in the cavity and transmitted wavefront slopes in optics in the cavity. Slopes will cause the rays not to be incident normally at the test and reference surfaces. Because of aberrations in the imaging system of the interferometer, the ray which is traveling at an angle may traverse a different optical path length than the ray which was normally incident. An OPD error results.

4.1 Source wavefront aberrations

For equal path interferometers, the source wavefront errors in the two arms of the interferometer are the same and the only effect is to distort the interferogram image; generally a negligible effect. For unequal path interferometers, the source wavefront changes shape over the length of the cavity, so that the reference and test wavefronts are no longer the same shape when they interfere. A simple example is de-collimation of the wavefront for a plano cavity (see Figure 3a) which introduces power into the interferogram. Similar effects are seen for higher order aberrations (see Figure 3b) and the effect is worse with greater wavefront slope. Ray-mapping errors due to source wavefront aberrations can be corrected for by reference measurement and subtraction, though care must be taken to maintain the cavity length. For spherical parts this requires a reference of the same radius as the part under test. For example, if a convex surface with a 3 mm radius of curvature (e.g., a ball bearing) is being tested, a reference measurement with a 25 mm radius concave reference will not properly subtract out the systematic ray-mapping errors. Specific examples of the MARK IV*xp* interferometer are shown in table 2.

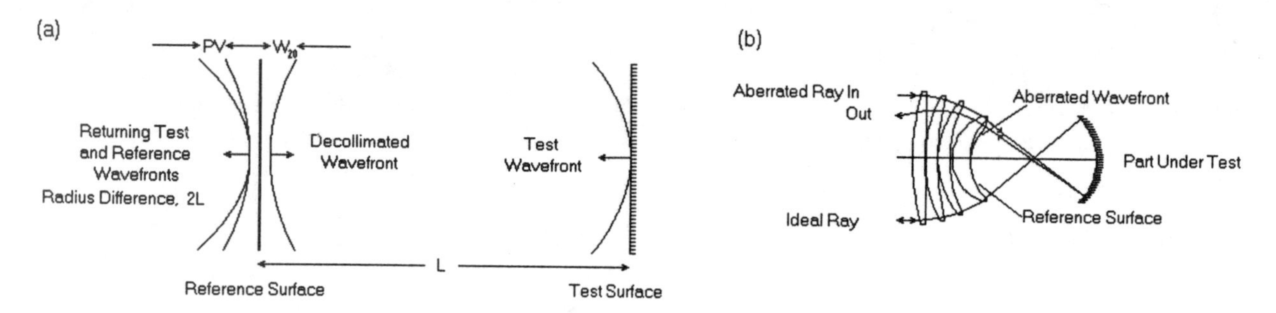

FIGURE 3. RAY-MAPPING ERROR DUE TO SOURCE WAVEFRONT ERRORS. (A) DE-COLLIMATED WAVEFRONT IN A PLANO CAVITY. (B) WAVEFRONT ABERRATIONS FROM A TRANSMISSION SPHERE (DIVERGER LENS). THE ERROR IS CAVITY LENGTH/RADIUS DEPENDENT. THOUGH SHOWN AS FIZEAU CAVITIES FOR CONVENIENCE, THE ERROR IS FOUND IN ALL UNEQUAL PATH INTERFEROMETERS.

Aperture	Cavity Length/Part Radius	PV error (λ)	rms error (λ)
102 mm	1 m	0.00012	0.00004
102 mm	15 m	0.012	0.0032
90 mm	1 m	0.00004	0.00001
90 mm	15 m	0.0036	0.00078
f/0.75	50 mm, concave	0.00042	0.00013
f/1.0	50 mm, concave	0.00014	0.00004

TABLE 3. PV RAY-MAPPING ERRORS DUE TO SOURCE WAVEFRONT ABERRATIONS FOR THE MARK IV*XP* WITH 102 MM TRANSMISSION FLAT AND F/0.75 TRANSMISSION SPHERE. NOTE CAVITY LENGTH DEPENDENCE (PLANO CONFIGURATIONS) AND APERTURE DEPENDENCE.

4.2 Surface slope errors

Slopes on the part cause the ray not to retro-reflect, directing the test ray away from the center of the imaging system entrance pupil (see Figure 5). The resulting OPD is simply the aberration of the imaging system which naturally becomes larger as the pupil is opened up, i.e., as the slope error becomes larger. Correction of the imaging system is fundamental to reducing this type of error because test surface slope errors are inseparable from the measurement. For spherical testing the magnitudes of ray-mapping errors are affected by the relative radii of the reference and test surfaces. If the radii differ greatly, small surface slope or source wavefront errors can yield sizeable measurement errors. Ray-mapping errors of this type are the primary source of error, sometimes several waves in magnitude, when the returning testing wavefront is aspheric. As a rule, high fringe densities are a good indicator that ray-mapping errors should be considered.

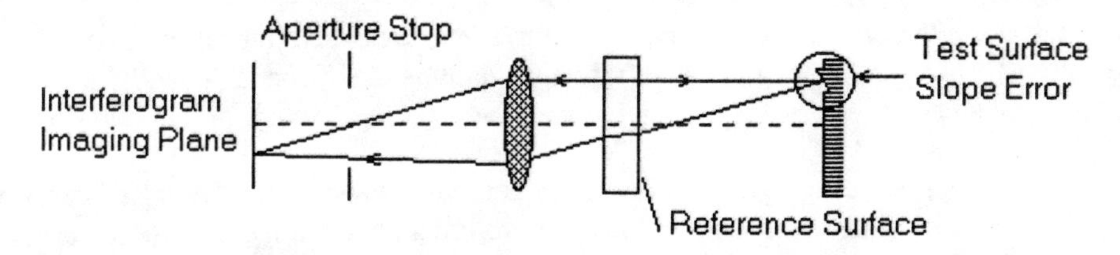

FIGURE 4. TEST SURFACE SLOPE ERRORS CAUSE THE TEST RAY TO RETURN ALONG A DIFFERENT PATH TO THE IMAGE PLANE. THIS IS SHOWN FOR A FIZEAU INTERFEROMETER.

Slope errors in the part have "instantaneous" or "cumulative" effects. Instantaneous effects are due to high frequency features like digs or diamond turning marks. These show the maximum OPD error at a small region against an otherwise perfect background. Cumulative effects are the variation in the instantaneous effects due to continuous changes in slope, e.g., tilt, power, mild aspheres.

Ray-mapping errors are of concern in any of the following cases (potential magnitudes are shown):

aspheric test wavefronts (high fringe density) (> 1λ)
high frequency surface slope errors (> 0.1λ), machined surfaces - "jagged" fringes
fast spherical diverger lenses (> 0.01λ)
long cavity lengths (> 0.01λ), air turbulence and vibration errors will usually be much worse

As with distortion, any part smaller than the interferometer aperture should be centered in the aperture to minimize ray-mapping errors.

5. ELECTRONICS ERRORS

If all optical measurement errors and on-site errors are eliminated, accuracy and precision are still limited by electronics errors in phase modulation and measurement. In fact, the instrumental precision of a phase measuring interferometer is typically determined by the electronics errors. Because of the prevalence of phase measuring systems and the higher accuracy of such systems, fringe detection is not discussed here.

The literature abounds with discussions of phase modulation/calculation[5,11-14]. Summarizing from this literature and from work at our facility, there are three important error sources in phase measuring interferometry: phase modulation calibration and linearity, digitization noise and linearity, and the phase calculation algorithm. The robustness of the phase calculation algorithm can make up for a number for shortcomings in the phase modulation. Five bucket algorithms are generally robust enough to provide instrumental repeatability well below the accuracy of the instrument accuracy. Digitization noise can be reduced by averaging data sets and/or intensity frames. Digitization linearity is strictly a function of design. The Zygo MARK IV*xp* has an electronic noise component of 0.00038λ PV, 0.000004λ rms. This includes detector noise, A/D noise and linearity, phase modulation linearity and calibration, and phase calculation errors. The phase modulation errors can be larger when using large and heavy transmission spheres.

6. ON-SITE ERRORS

On-site errors are those attributed to environmental sources and operator technique. While not part of the instrumental accuracy and precision, most high accuracy testing is limited by on-site errors. These errors can be corrected or at least identified.

6.1 Alignment

Rays reflect normally from both the reference and test surfaces in an ideal interferometer. If both surfaces are tilted equally, a nulled fringe pattern is maintained, but the reference and test wavefronts are sheared. For illumination wavefronts with spherical aberration, this will produce some degree of coma in the interferogram. This is an operator induced cumulative ray-mapping error.

Mis-alignment of an optic under test introduces errors. For example, tilting a lens measured in transmission will induce third and fifth order coma and astigmatism. Lens assemblies with reference datums may be mechanically or interferometrically aligned to reduce such errors.

Focus of the part under test is also an alignment issue[15]. Diffraction from unfocussed apertures adds significant errors to the measurement, usually affecting the PV statistic more than the rms or third order aberrations. In extreme cases, a λ/10 PV part can be measured in the range of λ/4 PV if thc focus is not properly adjusted. If long depth of focus is required, e.g., testing of laser rods in transmission, the interferometer can be modified to stop down the imaging system, sacrificing spatial resolution of the imaging system for depth of focus. It is also possible to exclude regions with diffraction from the measurement through software, though this means some to the part is not being measured.

6.2 Environment

Environmental errors include vibration, air turbulence, thermal gradients and mechanical drift. A rule of thumb for these errors is that if you can see it in the video image, then so can the interferometer.

Vibration is mostly seen as a uniform oscillation in the cavity length, though sometimes it can be non-uniform, e.g., pivoting or flexing fold mirror. Video based systems suffer most from the 1 - 60 Hz band. Vibration causes a phase error which appears as ripple in the phase data at twice the fringe frequency. Re-stated, if the interferogram has n straight fringes, the phase map will have 2n straight ripples. If the fringes are curved, the ripples will curve. Typical optical shop environments can create ripple of λ/30 PV. Sources of vibration are many, but the prime offenders are the air conditioning system (30 Hz) and nearby heavy machinery. Averaging of phase data sets will reduce the effects of vibration. Averaging of intensity data frames will lead to poor fringe contrast and data dropouts. Some phase measurement techniques have been developed specifically to reduce vibration effects[16,17].

There are two simple methods of gauging the magnitude of vibration errors. First, any "fluttering" of the intensity in a well nulled cavity is due to vibration. As a general rule, if you can see it, so can the interferometer. Second, introduce between five and ten fringes of tilt. Then acquire data, subtract all aberrations through third order and verify the presence of ripple from a 3D plot. PV and rms errors of the residual are due to vibration.

Air turbulence is easily observed as "swirling" of the fringes in a well nulled cavity. Covering the interferometer cavity to restrict air flow will reduce turbulence effects. This may have a negative side effect in air stratification in the cavity for horizontal systems. Turbulence has both random and 1/f noise components. To a limited extent, averaging will reduce its effects in phase measuring interferometry.

Thermal gradients can take two forms. In the case of air stratification, the measurements will have a cylindrical component, with the cylinder axis parallel to the table. This is mostly a problem in large aperture plano testing (> 300 mm diameter). Thermal gradients in the part under test or the transmission element or any other optical elements in the cavity will stress the element and produce cavity errors. The only cure for this is to allow the system to reach thermal equilibrium before any high accuracy testing.

Tooling can also produce errors. Stresses on optical elements can induce homogeneity errors and figure errors. Sag or drift in mounts during the measurement process can wreak havoc with phase modulation calibration. The cavity and instrumentation should be allowed to reach mechanical equilibrium.

On-site errors combine with electronics errors to determine the experimental precision. This is evaluated by taking two measurements and analyzing the difference. Subtraction of a reference plane/sphere may be necessary if mechanical drift is not to be considered. Repeating this process yields a mean and standard deviation of the PV and rms precision (Note - PV precision is not as useful or reliable a measure as rms precision). For example, a Zygo MARK IV*xp* yielded a mean rms precision of 0.00032λ and standard deviation of 0.00021λ over 100 measurement pairs each consisting of 10 averaged phase maps all acquired during an evening. This indicates 0.00032λ rms precision (50% confidence) and 0.00094λ rms precision (98% confidence, single sided distribution). The same system yielded 0.0016λ rms precision (50%) and 0.0028λ rms precision (98%) during a high foot traffic time of day. If great care is taken to isolate the system from all environmental interference, including operator intervention, the experimental precision will approach the instrumental precision.

7. CONCLUSION

Most optical testing is relative testing and as such is limited by cavity errors. For testing of commercial grade optics, commercially available references (λ/20 - λ/40 PV) are adequate for qualification. Custom optical elements and systems may require better accuracy. Absolute determination of the measurement accuracy requires an absolutely calibrated test surface. This is a cumbersome process. Absolute testing methods and/or reference measurements for subtracting system error can improve accuracy, but the metrologist must now know the magnitude of the secondary errors (distortion, ray-mapping and precision) which limit the measurement accuracy. On-site errors must be measured and, if possible, eliminated for high accuracy testing.

8. ACKNOWLEDGEMENTS

The author wishes to thank George Hunter, Peter Pfluke, Bruce Truax and Bob Smythe for helpful discussions and assistance.

9. REFERENCES

1. Eisenhart, C., "Realistic Evaluation of the Precision and Accuracy of Instrument Calibration Systems", NBS spec. pub. 300, **1**, p.21, 1969

2. Dorsey, N.E. and Eisenhart, C., "On Absolute Measurement", NBS spec. pub. 300, **1**, p.49, 1969

3. Schulz, G. and Schwider, J., "Interferometric testing of smooth surfaces", Progress in Optics, **XIII**, E. Wolf ed., North-Holland, 1976

4. Murty, M.V.R.K., "Newton, Fizeau and Haidinger Interferometers", Optical Shop Testing, D. Malacara ed., Wiley, 1978

5. Bruning, J.H., "Fringe Scanning Interferometers", Optical Shop Testing, D. Malacara ed., Wiley, 1978

6. Truax, B. E., "Absolute testing of spherical surfaces", *SPIE Proc.* **966**, 1988

7. Parks, R., "Removal of test optics errors", SPIE Proc. **153**, 1978

8. Fritz, B., "Absolute calibration of an optical flat", *Optical Engineering*, **23**, p.379, 1984

9. Selberg, L. A., "Interferometer accuracy and precision", *SPIE Proc.* **749**, 1987

10. Smith, W.J., Modern Optical Engineering, p.269, McGraw-Hill, 1966

11. Cheng, Y. Y. and Wyant, J.C., "Phase shifter calibration in phase-shifting interferometry", *Applied Optics*, **24**, p.3049, 1985

12. Moore, R. C. and Slaymaker, F. H., "Direct measurement of phase in a spherical-wave Fizeau interferometer", *Applied Optics*, **19**, p. 2196, 1980

13. Schwider, J., et. al., "Digital wave-front measuring interferometry: some systematic error sources", *Applied Optics*, **22**, p.3421, 1983

14. Kinnstaetter, K., et. al., "Accuracy of phase shifting interferometry", *Applied Optics*, **27**, 24, p. 5082, 1988

15. Malacara, D. and Menchaca, C., *SPIE Proc.* **540**, 1985

16. Wizinowich, P. L., "Phase shifting interferometry in the presence of vibration: a new algorithm and system", *Applied Optics*, **29**, p. 3271, 1990

17. Koliopolous, C. L., "Avoiding phase-measuring interferometry's pitfalls", *Photonics Spectra*, p. 169 October, 1988

Optical Testing with Wavelength Scanning Interferometer

Katsuyuki Okada and Jumpei Tsujiuchi

Department of Image Science & Engineering
Faculty of Engineering
Chiba University
1-33 Yayoi-cho, Chiba 260, Japan

Abstract

Wavelength change in a unbalanced interferometer makes phase change in interferograms, and this phenomenon can be used for the purpose of phase shifting interferometry. As an example of measurements, an optical plane parallel plate is tested, and the simultaneous measurement of both front and rear surfaces together with inhomogeneity of refractive index can be carried out. A new method for calculating phase distribution is developed under the condition where nonlinearity of tuning wavelength and fluctuation of output power take place.

1. Introduction

If the wavelength of a light source changes in a unbalanced interferometer, in which the testing and the reference beams have different path lengths, the phase difference between two beams changes. In such a case, let λ be the wavelength, $\Delta\lambda$ the amount of wavelength change and ΔL the path difference between the two beams, the phase change δ in the interferogram becomes with $k = 2\pi/\lambda$

$$\delta = \frac{2\pi\,\Delta L\,\Delta\lambda}{\lambda^2} = \frac{k\Delta L\Delta\lambda}{\lambda}. \tag{1}$$

So, if a series of interferograms are obtained by successive change of wavelength, it is possible to apply a technique of phase shifting interferometry without any mechanical movement, and this will give the advantage of high precision measurement with simple mechanism [1-4].

2. Measurement of Plane Parallel Plates

This technique is applied to the measurement of optical plane parallel plates. Fig. 1 shows a Twyman-Green interferometer with unbalanced beams, and a plane parallel plate S is put in the testing beam. A tunable laser diode LD is employed as a light source, a parallel beam from a collimator L_1 is divided into two by a beam splitter BS, and the divided beams go to the reference mirror M_1 and to the testing mirror M_2. An interferogram produced by the reference beam coming from M_1 and the testing beam coming from M_2 and passing through S can be observed in a plane D through an imaging lens L_2.

In such a measurement, the sample S has usually no anti-reflecting coating, and wavefronts reflected from front and rear surfaces of S appear together with those from mirrors M_1 and M_2, and these wavefronts interfere with each other. So, many interference fringes are superimposed and the resultant interferogram is too complicated to analyze by conventional fringe analyzing techniques.

However, as optical path differences between two wavefronts producing individual interferograms are not the same, the individual interferograms will move in different ways if the wavelength changes, and they can be analyzed by separating each other. In addition, since these interferograms include not only the wavefront passing through the sample S but also reflected wavefronts from both front and rear surfaces of S, the fringe analysis will make clear all the necessary parameters to evaluate the sample, i.e. surface shape of front and rear surfaces, unevenness of the thickness and inhomogeneity of refractive index [5,6].

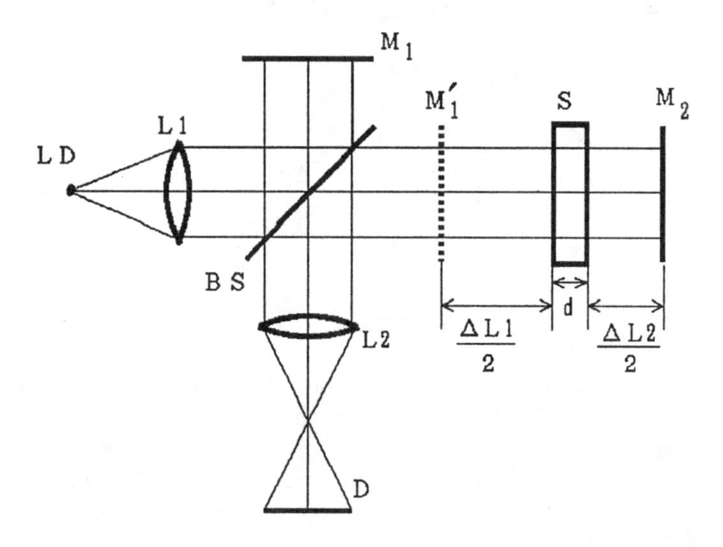

Fig. 1 Unbalanced Twyman-Green interferometer [5]

The resultant interferogram in this case is a superposition of the following six individual interferograms, i.e. 1) interferogram of wavefronts reflected from M_1 and M_2, 2) that of M_1 and the rear surface of S, 3) that of M_1 and the front surface of S, 4) that of the front and the rear surfaces of S, 5) that of M_2 and the front surface of S, and finally 6) that of M_2 and the rear surface of S.

If the wavelength of the laser diode is changed by every $\Delta\lambda$, the intensity distribution of the m-th interferogram with a wavelength change of $m\cdot\Delta\lambda$ becomes

$$\begin{aligned} I^{(m)}(x) = I_0 &+ I_1\cos[\phi_1(x)+\phi_2(x)+\phi_3(x)+mk(\Delta\lambda/\lambda)(\Delta L_1+2n_0d_0+\Delta L_2)] \\ &+ I_2\cos[\phi_1(x)+\phi_2(x)+mk(\Delta\lambda/\lambda)(\Delta L_1+2n_0d_0)] \\ &+ I_3\cos[\phi_1(x)+mk(\Delta\lambda/\lambda)\Delta L_1] \\ &+ I_4\cos[\phi_2(x)+mk(\Delta\lambda/\lambda)2n_0d_0] \\ &+ I_5\cos[\phi_2(x)+\phi_3(x)+mk(\Delta\lambda/\lambda)(2n_0d_0+\Delta L_2)] \\ &+ I_6\cos[\phi_3(x)+mk(\Delta\lambda/\lambda)\Delta L_2]. \end{aligned} \tag{2}$$

Terms from the 1st to the 6th lines of equation (2) correspond respectively to the above-mentioned six individual interferograms 1) - 6), where δL_1 is the optical path difference between M_1 and the front surface of S, δL_2 is that of M_2 and the rear surface of S, d_0 is the average thickness of S, and x is the one-dimensional expression of the coordinate in the surface of S. I_0, I_1, I_3, I_4, I_5 and I_6 are constants due to the output power of the laser diode and the reflectivity of the surfaces, and

$$\begin{aligned} \phi_1(x) &= 2ka(x) \\ \phi_2(x) &= 2kn(x)[b(x)-a(x)] \\ \phi_3(x) &= 2k[c(x)-b(x)] \\ \phi_4(x) &= 2kc(x) \end{aligned} \tag{3}$$

are phase distributions, in which $a(x)$ and $b(x)$ are functions expressing the shape of front and rear surfaces of S, and $c(x)$ is the shape of M_2.

The phase distributions can be obtained by solving equation (2). If the interval of wavelength change $\Delta\lambda$ is constant and the output power of the laser does not change during the measurement, we have 13 unknowns; i.e. I_0, $I_1\cos(\phi_1+\phi_2+\phi_3)$, $I_1\sin(\phi_1+\phi_2+\phi_3)$, $I_2\cos(\phi_1+\phi_2)$, $I_2\sin(\phi_1+\phi_2)$, $I_3\cos\phi_1$, $I_3\sin\phi_1$, $I_4\cos\phi_2$, $I_4\sin\phi_2$, $I_5\cos(\phi_2+\phi_3)$, $I_5\sin(\phi_2+\phi_3)$, $I_6\cos\phi_3$, and $I_6\sin\phi_3$, and equation (2) is linear with respect to these unknowns. So, it is possible to obtain these unknowns by using the least square fitting of equation (2) with a series of interferograms which are obtained by changing the wavelength with known parameters such as λ, $\Delta\lambda$, δL_1, δL_2, n_0 and d_0.

The least square fitting is carried out as follows; first

$$Q = \sum_{m=1}^{M} [I(x,m) - I^{(m)}(x)]^2 \tag{4}$$

is calculated for every x, where $I(x,m)$ is the measured intensity of the m-th interferogram and $I^{(m)}(x)$ is given by equation (2), and M is the total number of interferograms obtained by changing the wavelength. Then, differentiate Q with respect to the above mentioned unknowns, and set them equal to 0, thus a set of simultaneous equations are obtained and they can be solved by computer. Phase distributions ϕ_1, ϕ_2 and ϕ_3 are obtained by

$$\begin{aligned} \phi_1(x) &= \tan^{-1}[I_3\sin\phi_1(x)/I_3\cos\phi_1(x)] \\ \phi_2(x) &= \tan^{-1}[I_4\sin\phi_2(x)/I_4\cos\phi_2(x)] \\ \phi_3(x) &= \tan^{-1}[I_6\sin\phi_3(x)/I_6\cos\phi_3(x)]. \end{aligned} \tag{5}$$

ϕ_4 is the phase distribution of M_2, and measured directly from the interferometer by removing S from the testing beam. Finally, we have

$$\begin{aligned} a(x) &= \phi_1(x)/2k \\ b(x) &= [\phi_3(x) - \phi_4(x)]/2k \\ c(x) &= \phi_4(x)/2k \\ d(x) &= d_0 + a(x) - b(x) = d_0 + [\phi_1(x) - \phi_3(x) + \phi_4(x)]/2k \\ n(x) &= [n_0 d_0 + \phi_2(x)/2k]/d(x). \end{aligned} \tag{6}$$

This method is very suited for testing optical plane parallel plates such as filters or windows, and all necessary parameters for evaluating the plate can be obtained just in one measurement. In the experiment, a tunable laser diode with $\lambda = 795$ nm is used, and 60 interferograms are taken by a 256 × 256 pixels 8 bits CCD camera. The average driving current is 65 mA and the total amount of current change is about 5 mA which corresponds to a wavelength change of 0.05 nm, this will satisfy the condition of linear wavelength tuning and constant output power of the laser diode. A glass plate of $d_0 = 4.80$ mm and $n_0 = 1.533$ is measured, and the RMS error of the measurement is estimated at 1/60 wavelength for the shape and 10^{-5} for the refractive index [5,7]. The influence of the dispersion of the sample can be compensated easily by changing d_0 [5].

3. Measurement with Nonlinear Wavelength Tuning

Sometimes, however, a laser diode has nonlinearity of wavelength tuning in such a way that the change of driving current is not proportional to the change of wavelength and it may cause the change of output power and coherence as well. In such a case, the interval of wavelength change $\Delta\lambda$ is not constant any more, and it is important to know the exact amount of wavelength change $m\cdot\Delta\lambda$ or the resultant phase change δ. For that purpose, a high precision spectrometer or an additional interferometer is needed to evaluate $\Delta\lambda$ or δ, but it seems not practical. In this connection, a new fringe analyzing technique is developed to find both the phase shift δ and the phase distribution ϕ. The change of output power of the laser diode is also taken into consideration.

A unbalanced Twyman-Green interferometer for measuring the shape of the testing mirror M_2 is taken as an example for simplicity, and the intensity distribution of the m-th interferogram is given by

$$I^{(m)}(x) = I_0 + I_1 \cos[\phi(x) + \delta(m)], \tag{7}$$

where $\delta(m)$ is the amount of phase shift in the m-th interferogram and is not known as well as the phase distribution $\phi(x)$ of the testing mirror. To obtain these values, the least square fitting is applied, and

$$Q = \sum_{m=1}^{M} \sum_{x=1}^{N} [I^{(m)}(x) - I(x,m)]^2 \tag{8}$$

is calculated for every x and m instead of equation (2) to find unknowns $\phi(x)$, $\delta(m)$, I_0 and I_1.

However, two problems exist in this case: One is that both ϕ and δ are in the argument of cosine and the equation is not linear with respect to these unknowns. The other one is that the unknowns I_0, I_1 and $\phi(x)$ should be solved for all sampling points x and $\delta(m)$ for all interferograms, i.e. the total number of unknowns is $3N+M-1$, and this is too many to solve with the conventional method. Accordingly, a new iterative method is developed for solving them [8].

In fact, if we take a point x on the surface of the sample, the value of $\phi(x)$ is constant in all the interferograms for m=1 to M, and so far as the m-th interferogram is concerned values of I_0, I_1 and $\delta(m)$ are constant. This means that the relation between δ and ϕ would be weak, and the calculation can be separated into two stages. First, the value of $\phi(x)$ is calculated by the same method as the previous section under the assumption that the value of $\delta(m)$ is known, then more reasonable value of $\delta(m)$ can be calculated by using just obtained $\phi(x)$ in the m-th interferogram. Repeating these iterative calculations, $\delta(m)$ and $\phi(x)$ converge to the correct values respectively. In addition, if there is the change of output power of laser diode as well as the change of coherence, the intensity of interferogram can be expressed, by separating both I_0 and I_1 into two components, as follows;

$$I^{(m)\prime}(x) = I_0(x) I_0'(m) + I_1(x) I_1'(m) \cos[\phi(x) + \delta(m)] \tag{9}$$

then the least square fitting starts from calculating

$$\begin{aligned} Q &= \sum_{m=1}^{M} \left[I^{(m)\prime}(x) - I(x,m)\right]^2 \\ &= \sum_{m=1}^{M} \left[I_0'(m) I_0(x) + I_1'(m)\cos\delta(m) I_1(x)\cos\phi(x) - I_1'(m)\sin\delta(m) I_1(x)\sin\phi(x) - I(x,m)\right]^2, \end{aligned} \tag{10}$$

and unknowns to be found are $I_0(x)$, $I_1(x)\cos\phi(x)$ and $I_1(x)\sin\phi(x)$ for every x. After getting these unknowns with an appropriate initial guess of $\delta(m)$, we calculate

$$\begin{aligned} Q &= \sum_{x=1}^{N} \left[I^{(m)\prime}(x) - I(x,m)\right]^2 \\ &= \sum_{x=1}^{N} \left[I_0(x) I_0'(m) + I_1(x)\cos\phi(x) I_1'(m)\cos\delta(m) - I_1(x)\sin\phi(x) I_1'(m)\sin\delta(m) - I(x,m)\right]^2, \end{aligned} \tag{11}$$

and obtain $\delta(m)$, $I_0'(m)$ and $I_1'(m)$ for every m. Repeating these calculations, $\phi(x)$ and $\delta(m)$ converge to the final and correct values.

These iterative calculations can conclude correct values of both phase distribution $\phi(x)$ and scanning phase shift $\delta(m)$ in the same time, and very high precision measurement can be carried out. So, even if a laser diode has nonlinearity of wavelength scanning, the interferometer will function with satisfactory accuracy. According to computer simulations and practical experiments, the iteration converges only in a few steps, and the number of sampling points to determine the phase shift $\delta(m)$ in equation (11) can be reduced down to only a few [8]. This fact shows that the time of calculation can be significantly reduced.

4. Conclusion

A wavelength scanning interferometer using a tunable laser diode as a light source is proposed for optical testing, and gives an interesting method for measuring optical parallel plates. This method is very simple, and phase shifting can be made only by changing the driving current of the tunable laser diode without any mechanical movement. Even if the laser diode has nonlinearity in tuning wavelength and fluctuation of output power, a new iterative calculation will give satisfactory result in a short time.

References

[1] Y. Ishii, J. Chen and K. Murata: "Digital phase-measuring interferometry with a tunable laser diode", Opt. Lett. **12**, 233-235 (1987).
[2] A. Valentin, C. Nicholas, L. Henry and A. W. Mantz: "Tunable diode laser control by a stepping Michelson interferometer", Appl. Opt. **26**, 41-46 (1987).
[3] K. Tatsuno and Y. Tsunoda: "Diode laser direct modulation heterodyne interferometer", Appl. Opt. **26**, 37-40 (1987).
[4] J. Chen, Y. Ishii and M. Murata: "Heterodyne interferometry with a frequency-modulated laser diode", Appl. Opt. **27**, 124-128 (1988).
[5] K. Okada, H. Sakuta, T. Ose and J. Tsujiuchi: "Separate measurements of surface shapes and refractive index inhomogeneity of an optical element using tunable-source phase shifting interferometry", Appl. Opt. **29**, 3280-3285 (1990).
[6] K. Okada and J. Tsujiuchi: "Wavelength scanning interferometry for the measurement of both surface shapes and refractive index inhomogeneity", Proc. SPIE, vol. **1162**, 359-401 (1989).
[7] K. Okada, H. Sakuta, T. Ose and J. Tsujiuchi: "Error analysis on the phase calculation from superimposed interferograms generated by a wavelength scanning interferometer", Opt. Commun. **77**, 343-348 (1990).
[8] K. Okada, A. Sato and J. Tsujiuchi: "Simultaneous calculation of phase distribution and scanning phase shift in wavelength scanning interferometer", Opt. Commun. to be published.

DIGITAL TALBOT INTERFEROMETER

S.C. TAM*, D.E. SILVA** and H.L. WONG**
* School of MPE, Nanyang Technological Institute, Singapore 2263
** Polycore Optical Pte Ltd, Singapore 1334

ABSTRACT

A computer based automatic measuring instrument has been developed to compute the power distribution of progressive lenses. The measurement principle is based on Talbot interferometry in which the self image of a Ronchi grating is made to interfere with an analyzing grating of equal period. The fringe patterns produced are captured with a CCD camera and digitized with a Data Translation DT2851 frame grabber card. A C-langauge software package has been developed to perform fringe analysis using spin filtering, weighted averaging, and fringe centerline extraction with the binary derivative method. The power distribution is computed from the slope of the fringe centers. The instrument has been applied successfully to evaluate constant power and progressive power lenses.

1. INTRODUCTION

Optical power is the most fundamental property of a lens. Consequently many methods were developed over the past century to measure it. In the ophthalmic industry, the lensometer, or focimeter, routinely and accurately measures spherical and torical lenses. When measuring bifocal lenses, the lensometer is used twice; once for each focus. For progressive lenses, a complete characterization is done by repeatedly positioning the lens over the lensometer until the entire surface is measured. The process is very time consuming taking several minutes.

Nakano and Murata[1] used a shearing interferometer based on the Talbot effect to measure the power distribution of progressive lenses. Through a single observation of fringes, the entire lens is measured for power. This has a practical application for production and quality control of progressive lenses, as well as providing a quick method of comparing different progressives.

In this paper we present a novel and insightful explanation of the interferomenter, followed by a description of a software system that automates the reduction of fringes. The power distribution of the progressive lens is computed and displayed.

2. THEORY

A rather remarkable optical phenomenon was first reported by H. F. Talbot[2] over one hundred and fifty years ago. Briefly stated, this phenomenon that we call the "Talbot effect" concerns the formation of images when a certain class of structures like a grating are illuminated by spatially coherent light. For plane waves at normal incidence, the self-images repeat themselves at intervals deduced by Lord Raleigh[3] as $2d^2/\lambda$ (where d is the grating period and λ is the wavelength of the light). In Figure 1 we show the optical schematic of the Digital Talbot Interferometer. Talbot self images are formed of grating g_1 by the collimated light.

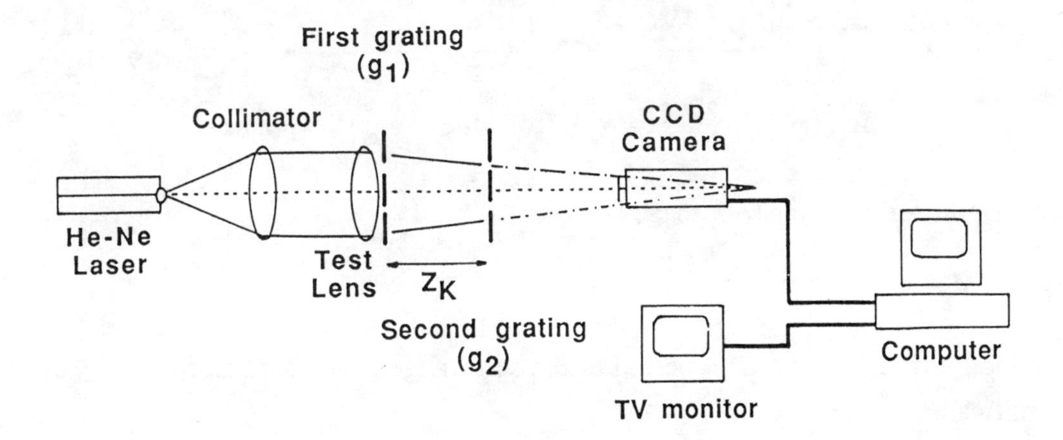

Fig. 1 Optical schematic of the digital Talbot interferometer

Cowley and Moody[4] have shown that the position of the self image planes of a grating illuminated by a spherical wavefront are given by,

$$\frac{1}{Z_K}+\frac{1}{Z_w}=\frac{1}{KZ_T} \qquad \ldots(1)$$

where Z_W = wavefront radius of the grating, Z_K = distance to the observed self-image, Z_T = first positive Talbot image distance = $2d^2/\lambda$, and $K = \pm\frac{1}{2}, \quad \pm 1, \quad \pm\frac{3}{2}, \ldots.$

From Figure 2, it can be seen geometrically that the ratio of the period of the first grating to that of its self Talbot image falling on the second grating is

$$\frac{d}{d'}=\frac{f}{(f-Z_K)}$$

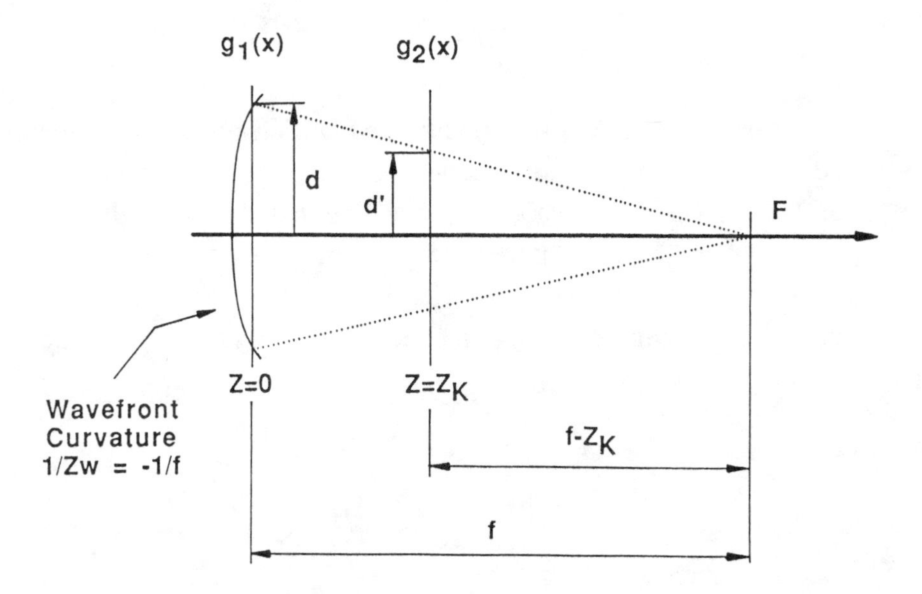

Fig. 2 A converging wavefront due to a positive lens

Therefore, the magnification of the self image is,

$$M = 1 + \left(\frac{Z_K}{Z_W}\right) = 1 - \left(\frac{Z_K}{f}\right) \qquad \ldots(2)$$

where d = period of first grating, d' = period of the self image of the first grating, f = focal length, and M = magnification factor.

For a lens touching the first grating, it can be assumed that Z_W is equal to the focal length of the lens. Hence, for Ronchi grating of period d its Talbot image will have a magnified period of Md. $M < 1$ for positive lens (because $Z_W < 0$, i.e. converging illumination) and $M > 1$ for negative lens ($Z_W > 0$, i.e. diverging illumination).

A second grating placed in one of the self image planes will create a moire pattern indicative of the curvature $1/Z_W$. Rogers[5] has used a diagram that simplifies the analysis. Grating effects are plotted in a reciprocal space similar to the Fourier frequency space. A grating of period d will be represented as a vector with a magnitude of $1/d$ and a direction perpendicular to the rulings, i.e. the direction lies along the diffraction pattern. From Figure 3, the fringe spacing for collimated light without test lens, P, is derived from the Rogers' diagram to be:

$$P = \frac{d}{\left[2\sin\left(\frac{\theta}{2}\right)\right]} \qquad \ldots(3)$$

where P = fringe spacing, d = grating period, and θ = angle between the two gratings.

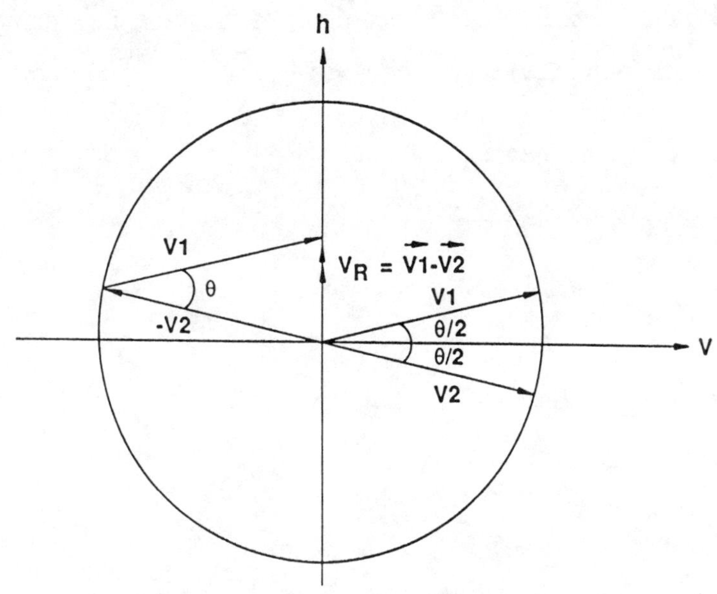

Fig. 3 A Rogers' diagram vector representation of gratings and fringe frequencies without a test lens

We consider a negative lens placed before grating g_1. The grating image period d' is magnified, $d' = Md$. Using the Rogers' diagram shown in Figure 4 (orientation adjusted), the vector magnitudes are V_1 for the Talbot image and V_2 for the second grating, respectively.

$$V_1 = \frac{1}{Md} \tag{4}$$

$$V_2 = \frac{1}{d} \tag{5}$$

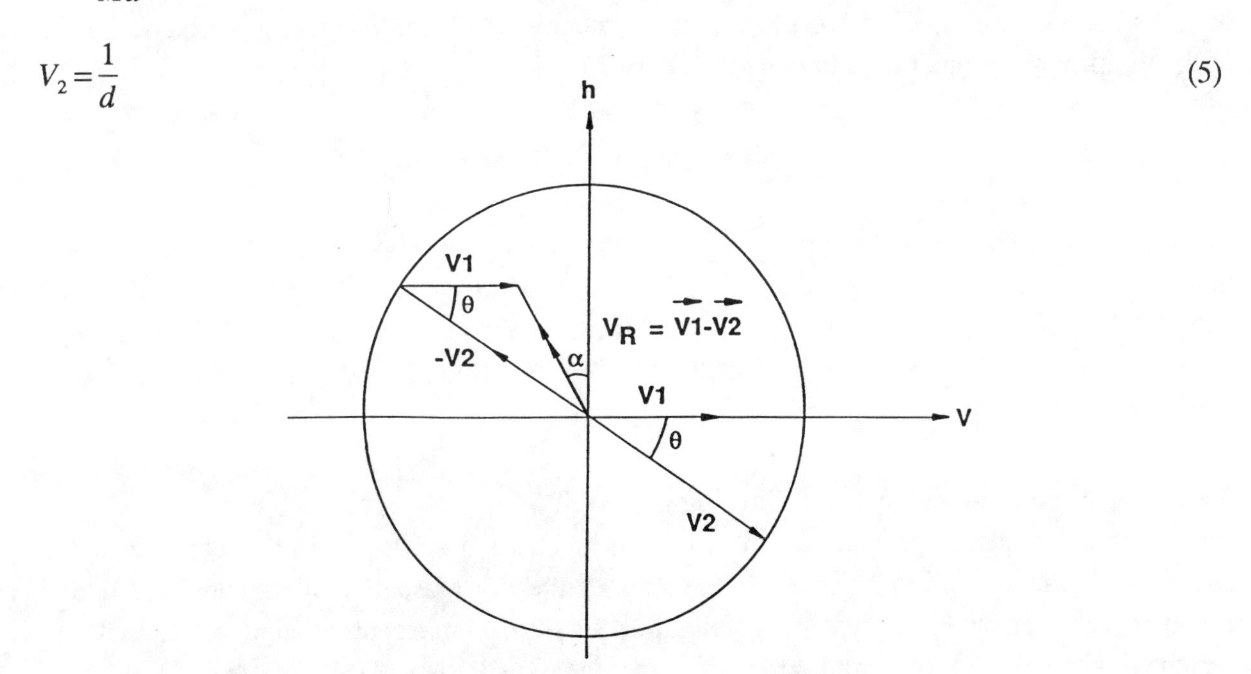

Fig. 4 Rogers' diagram for a negative test lens

The angle between $\vec{V}_1$ and $\vec{V}_2$, θ, is equal in magnitude to the angle between the gratings g_1 and g_2. The moire fringe vector (beat frequency), $\vec{V}_R$, is equal to $\vec{V}_1 - \vec{V}_2$. The angle α between $\vec{V}_R$ and the vertical axis is the angle of tilt of the moire fringes which is used as a measure of the focal length. Resolving the horizontal and vertical components,

$$V_{RH} = V_1 - V_2 \cos\theta = \frac{(1 - M\cos\theta)}{Md}$$

$$V_{RV} = V_2 \sin\theta = \frac{\sin\theta}{d}$$

The fringe rotation angle is given by

$$\tan\alpha = \frac{V_{RH}}{R_{RV}} = \frac{1 - M\cos\theta}{M\sin\theta}$$

Substituting M from Equation (2),

$$\tan\alpha = \left[\left(\frac{f}{f - Z_K}\right) - \cos\theta\right] / \sin\theta \qquad \ldots(6)$$

Substituting Z_K from the Cowley and Moodie equation,

$$\therefore \quad f = \frac{1}{(\sin\theta\tan\alpha + \cos\theta - 1)} \times \frac{Kd^2}{\lambda} \qquad \ldots(7)$$

Equations (6) and (7) are the two equations that can be used for the calculation of focal length. Equation (6) is a more general equation and requires 2 parameters, i.e. the tilt angle, θ, and the distance between the two gratings, Z_K. The latter quantity is computed from the grating period d, the wavelength of the light source λ, and the Talbot image order K.

3. DATA REDUCTION SOFTWARE

The image files captured and digitized using a Computar CCD camera and a Data Translation DT 2851 high-resolution frame grabber card are stored in binary format with gray level values of 0 to 255. The size of the image frame is 512 x 512 pixels, which requires a lot of memory space and long computational time. Hence, the size of the file is proportionally reduced to 200 x 200 pixels. The operation sequence of the software is shown in Figure 5.

3.1 Noise reduction using Spin Filtering and Weighted Averaging

The digitized moire fringe patterns are noisy due to laser speckles and electronic noise. Efforts to reduce the noise by optical hardware alone such as the use of spatial filters and laser line filters are found to be ineffective. Hence, two techniques of digital image processing, i.e. spin filtering[6] and weighted averaging, were employed to reduce the noise to a level acceptable for accurate data reduction.

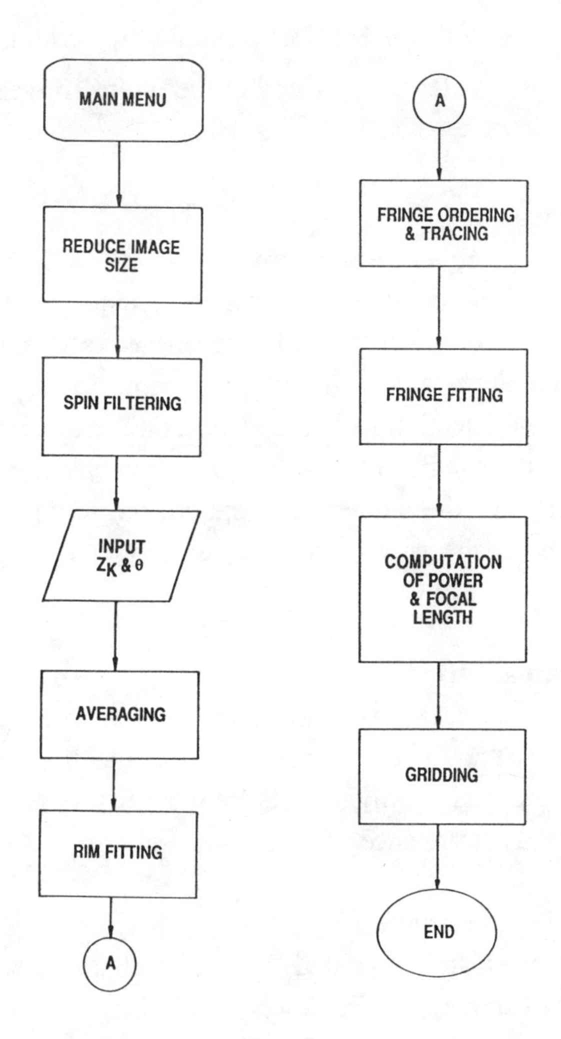

Fig. 5 Flowchart of power and focal length reduction software

Spin filtering exploits a property of the moire fringes in that the variation gradient of gray levels in neighboring points takes a minimum in the direction of the tangent of the fringe, a maximum in the direction normal to the fringe, and medium in other directions. Thus, the tangent of the fringes can be estimated from a knowledge of the gradients of the gray levels in the 8 spin directions. The average value of this filter line can well represent the original fringe feature of the current point. The result is not only the filtering of noise but also the retrieval of the fringe pattern without any blurring effect.

Weighted averaging is a technique that is very effective in smoothening noise while maintaining the location of the original peaks. However, the gray level magnitude is slightly reduced in the process. In this algorithm, weighted averaging is performed along the columns of the moire fringes. The weighted average of 5 pixels replaces the value of the current pixel.

3.2 Rim fitting

After the user has input the value of the distance between the two gratings, Z_K, and the relative tilt angle between the two gratings, α, a circular rim is fitted to the moire fringes to define the area of analysis. Only data within the rim is analyzed.

3.3 Fringe ordering and tracing

The moire fringes can be ordered either semi-automatically or fully automatically. In the semi-automic mode, the gray-level profile of the center column of the fringe pattern is displayed and the user is required to key in the coordinates of the peaks and number them sequentially. The software then traces or connects the peaks over the entire area within the fitted rim. In the fully automatic mode, the moire fringes are ordered using a peak detection technique based on the binary derivative method[6]. Slopes on opposing sides of the peak of a fringe have opposite signs. By detecting the transition of signs of slopes along the center row or column of the moire fringe pattern the peaks are located and numbered. The software then traces or connects the peaks over the entire area within the rim thereby reducing the fringes to lines.

3.4 Least squares polynomial fitting

The fringe lines are fitted to nonlinear curves using the least squares polynomial fitting method. By differentiating the equation of the fitted nonlinear curves, the tilt angle along each fringe can be computed. The software fits the data to polynomials from the first degree to the fifth degree, and selects the degree which gives the smallest variance.

3.5 Power and focal length computation

The slopes of the fitted fringes are computed by differentiating the polynomial equation of the respective fringes. This is substituted into Equation (6) to obtain the focal length and then the power. The focal length and power are computed in this manner for all the points along each fringe over the entire area defined by the rim.

3.6 Gridding

Gridding is done using the SURFER software produced by Golden Software, Inc. A regular spaced grid is generated over the entire rim using the inverse distance interpolation method. Data points are weighted with a weighting factor such that the influence of one data point on another decreases with distance from the point being estimated. For each interpolated grid value, a search area is defined. Only data points within the search area and closer than a specified search radius are used in the interpolation. The grid values are then "smoothed" using a smoothing matrix to weigh the nearest neighbors.

4. SYSTEM INTEGRATION AND TESTING

The Talbot Interferometer is assembled on an optical breadboard. The schematic representation of the system is shown in Figure 1. Two Ronchi gratings of pitch 7 lines/mm are used. The gratings are illuminated by an expanded collimated beam from a Melles Griot 5 mW linearly polarized 632.8 nm HeNe laser unit. Spatial filtering of the beam is performed with a 40X microscope objective and a 25 micron pin-hole. The beam is then collimated and expanded to a diameter of 50 mm using a laser collimator.

The relative tilt angle between the gratings is obtained by rotating the gratings in opposite directions until horizontal fringes appear. The fringe spacing is measured and the value of the angle of rotation between the two gratings is derived from Equation (3). The relative tilt angle between the two gratings is set to 0.5 degrees approximately.

The test lens is inserted immediately before the first grating and the spherical light wave that emerges from the test lens illuminates the gratings. The first grating is fixed; the second grating is moved along the z axis until a moire fringe pattern with good contrast is obtained.

The moire fringes are captured and digitized using a CCD camera and a high resolution frame grabber card residing in a 386 compatible computer. Five frames are captured and averaged to reduce noise. The digitized image is subsequently stored as gray level values between 0 to 255 in binary format for further reduction. The power distribution of the lens is obtained from the reduction C-program and presented graphically using the "SURFER" software.

A positive constant power lens and two progressive power lenses have been tested. Plate 1 shows the moire fringe pattern of a positive constant power lens. The measured power of the lens using a lensometer is +0.95D. The power obtained from the digital Talbot interferometer is +0.92D, a discrepancy of 3% approximately. Plate 2 shows the moire fringe pattern of a base 6 add 100 progressive front wafer. The power distribution displayed in orthographical and topographical formats are shown in Figure 6 and Figure 7 respectively.

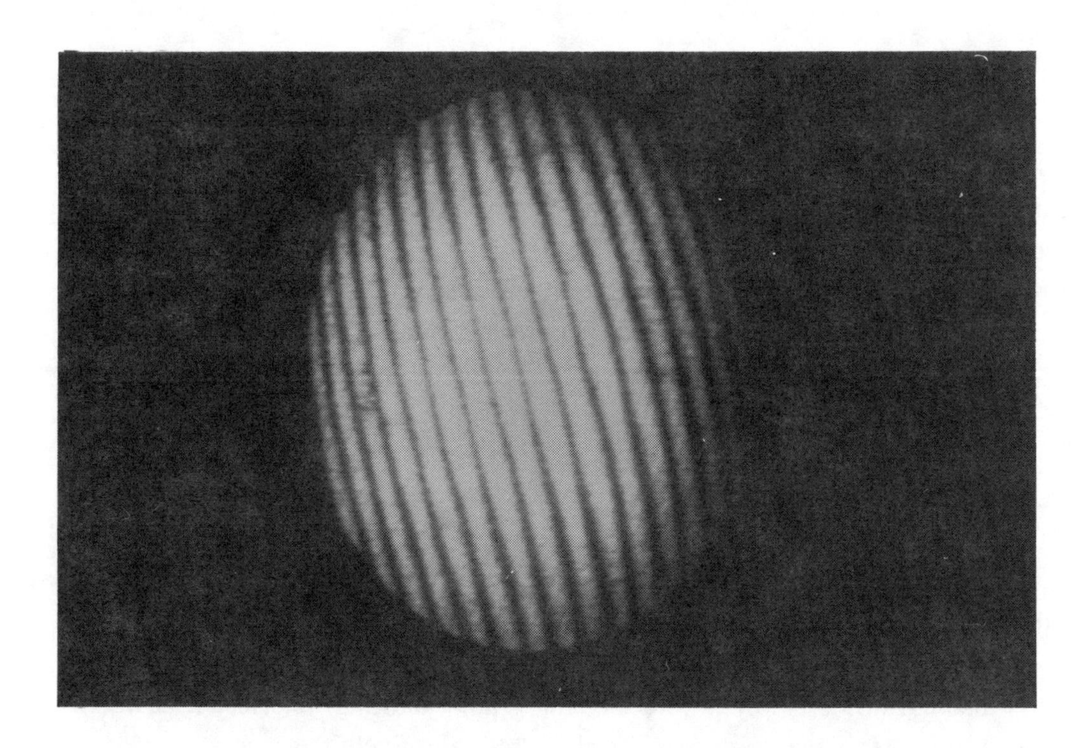

Plate 1: Moire fringe pattern of a + 0.95D constant power lens

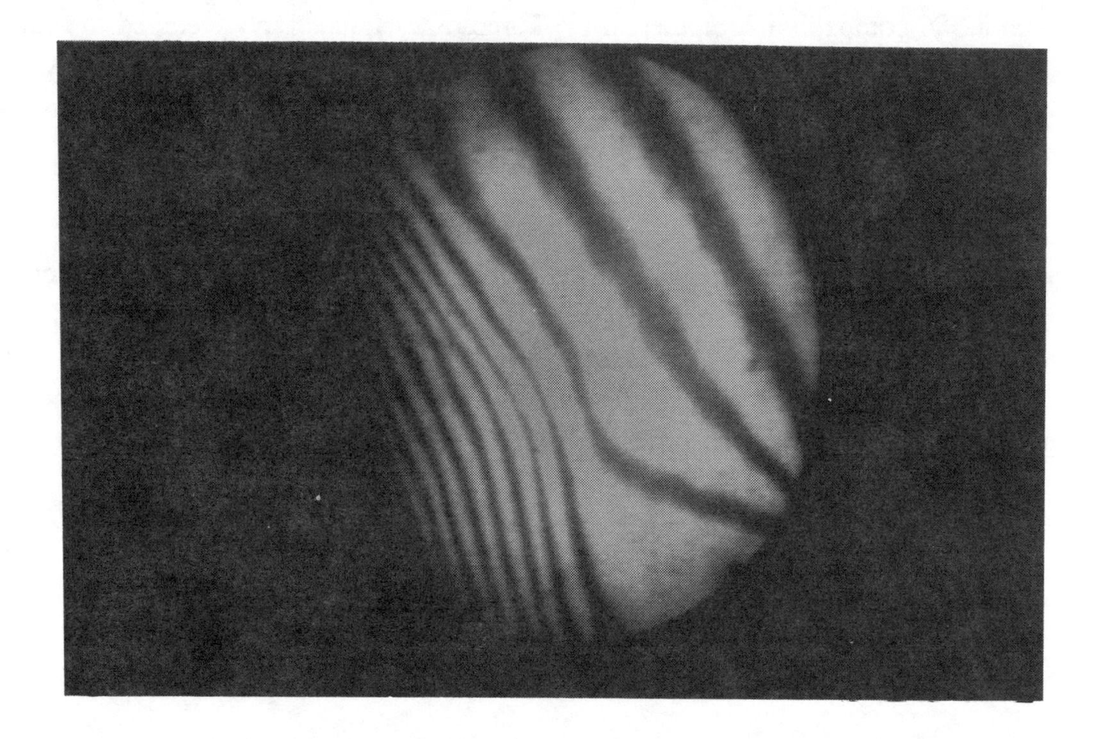

Plate 2: Moire fringe pattern of a base 6 add 100 progressive front wafer

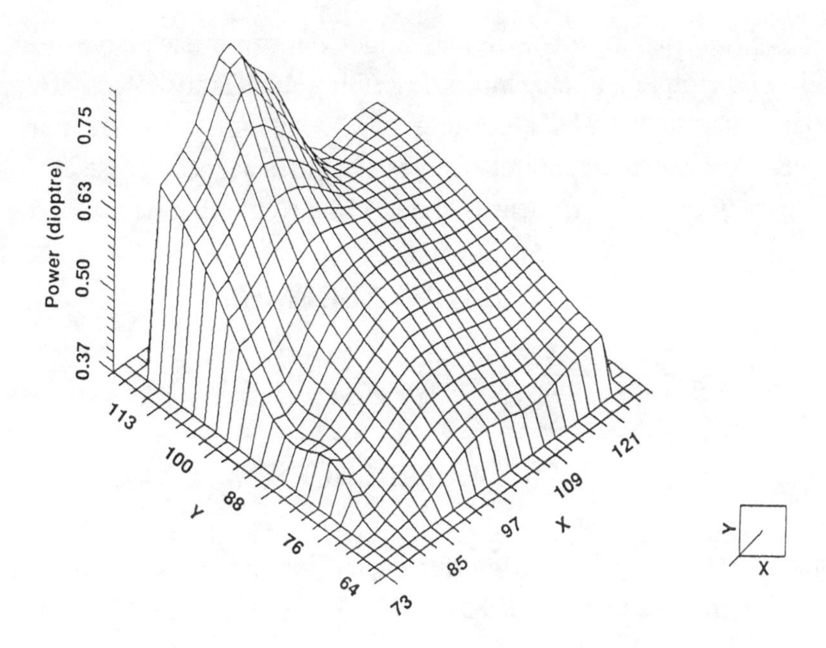

Fig. 6 An orthographical display of the power distribution of a base 6 add 100 progressive front wafer

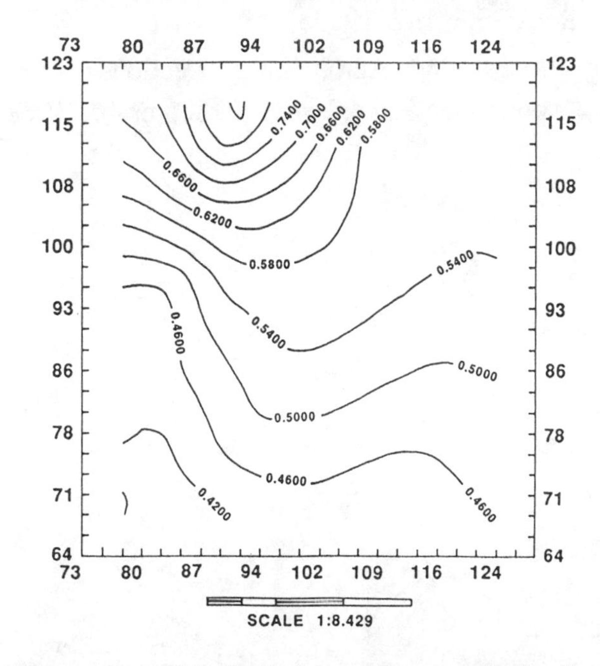

Fig. 7 A topographical display of the power distribution of a base 6 add 100 progressive front wafer

5. CONCLUSION

The theory, integration, and testing of a digital Talbot interferometer have been described. Rogers' diagrams are found to be simple and elegant in describing the Talbot effect. Hardware integration has been successfully implemented. The algorithm of a reduction software that evaluates the power distribution from the slopes of the captured fringes has also been discussed. Initial test results have indicated that the accuracy of the instrument is about 3%, a figure that has yet to be refined.

6. ACKNOWLEDGEMENT

We thank Ms Soh Meow Chng for typing this manuscript.

7. REFERENCES

1. Y. Nakano and K. Murata, "Talbot Interferometry for Measuring the Focal Length of a Lens", Applied Optics, 24, p.3162, October 1985.
2. H.F. Talbot, Phil. Mag., 9, p.401, 1836.
3. Lord Raleigh, Phil. Mag., 11, p.196, 1881.
4. J. Cowley and A. Moodie, Proc. Phys. Soc., B70, p.486, 1957.
5. G.L. Rogers, Proc. Phys. Soc., 73, p.142, 1959.
6. Q. Yu, "Spin Filtering Process and Automatic Extraction of Fringe Centrelines in Digital Interferometric Patterns", Applied Optics, 27, p.3782, September 1988.

SESSION 2A

Interferometry and Associated Techniques

Chair

Mike Driver
Avimo Singapore Pte. Ltd. (Singapore)

A High sensitivity interferometric technique for strain measurements

Arkady S. Voloshin* and Adel F. Bastawros**

* Department of Mechanical Engineering and Mechanics
Lehigh University, Bethlehem, Pa 18015, USA

** Homer Research Laboratories, Bethlehem Steel Corporation
Bethlehem, Pa 18016, USA

1. ABSTRACT

A high sensitivity strain measurement procedure that combines moiré interferometry and digital image processing, has been successfully implemented to determine thermally induced strains in electronic components. The technique is called Fractional Fringe Moiré Interferometry (FFMI). It produces whole field displacement information that are used to compute strains in a certain plane. Displacements in the submicron domain are detected with excellent spatial resolution over the area of interest. An example is presented here to illustrate the use of the technique to monitor thermally induced deformations in a specimen made from a plastic DIP device. The specimen was uniformly heated from room temperature to 90 °C, and the resulting moiré fringe patterns were recorded, analyzed using digital-image-processing and in plane displacements in the package were determined. Strain components were then computed by simple differentiation of the acquired displacement fields. Contour maps showing actual thermo/mechanical strain components in the device were constructed. Those maps can provide an excellent tool for strain analysis of microelectronic devices regardless of the structural complexity of the device.

2. BACKGROUND

Determination of strain levels in electronic packages is an essential requirement for addressing several strain related problems in packages. Severe thermal strains or thermal strain gradients may occur as a result of heating either during testing or processing of the package. Due to mismatch of the coefficients of thermal expansion of the different materials coexisting in a device, severe strain concentrations may occur thus leading to the ultimate failure of the device. Many analytical [1], and numerical [2] assessments of devices under such loading conditions were carried out in an effort to predict thermally induced deformations. However, the accuracy of the obtained results has been limited by the complexity of the device details, idealized material behavior, and many simplifying assumptions that had to be made to enable the analysis. This, of course, has limited the faithfulness of such models in representing behavior of actual packages under thermo/mechanical loading. They served, however, as good tools for qualitative assessments. On the other hand experimental determinations of strains in a package have not gone far beyond photoelastic modelling[3] and use of stress chips [4]. The latter gave information limited to the locations of the strain gages on the upper surface of the chip. No information is available by this approach about the remaining components of the package or even about the rest of the silicon chip itself. In addition, the reliability of data collected by this technique is largely affected by the absence of a proper calibration procedure for the strain gages.

It is the objective of this paper to introduce an ultra-sensitive interferometric technique that has been developed in our photomechanics laboratories, for in-situ full field strain measurements in electronic packages subjected to thermal loading. The technique used here is moiré interferometry enhanced by a digital-image-analysis system to evaluate displacements at a resolution that is much higher than currently available.

3. FRACTIONAL FRINGE MOIRÉ INTERFEROMETRY (FFMI)

Interferometric moiré has been recently introduced [5,6] as a sensitive full field deformation measuring technique. It is based on the formation of fringes by the coexistence of light wave fronts diffracted from a specimen grating of high frequency (e.g. 1200 lines/ mm). These specimen gratings are created on the specimen surface using a simple replication technique from a specially prepared mold. Two beams of coherent laser light illuminate the specimen grating obliquely from an angle $+\alpha$ and $-\alpha$ to create a virtual reference grating in front of the specimen. The creation of this virtual grating is due to the formation of walls of constructive and destructive interferences when two coherent beams intersect at an angle, Figure 1. The frequency f of this grating is given by

$$f = (2/\lambda) \sin \alpha \tag{1}$$

where λ, is the wavelength of the light used. The reference frequency f is typically chosen to be 2400 lines/ mm for a specimen grating frequency of 1200 lines/mm. This corresponds to a sensitivity of 0.417 μm per fringe order and a multiplication factor of two. In interferometric moiré the reference grating frequency is always twice the specimen grating frequency to be able to collect the first order diffractions by a camera. On the other hand, specimen gratings are produced by replication from photographic plate molds that already have the grating pattern of the correct frequency printed on them. The grating has a very thin layer of metallic film that has been vacuum deposited on its surface to make it reflective. This mold is then pressed lightly against the specimen while entrapping a thin layer of high temperature epoxy. When the epoxy is fully cured the mold is pried off the specimen leaving a reflective epoxy replica of the grating pattern on the specimen, Figure 2.

When the specimen is deformed fringes are formed by the interference of the virtual gratings with light diffracted from the specimen gratings. The displacement is related to the fringe value and the reference grating frequency by the following expression

$$U = (1/f)\, N \tag{2}$$

where U is the displacement component in the direction normal to the grating and N is the fringe value. If crossed gratings are used, two normal displacement components (U and V) may be monitored simultaneously.

Determination of the fringe value at a point and consequently the displacement component at that point is done using the digital image analyzer which records the moiré field by means of a CCD Video camera. Light intensities at different picture elements (pixels), as detected by the camera, are digitized in terms of 256 gray levels. This array of light intensities is used to compute deformations at all points of the field using a basic optical law [7] that relates light intensities in a moiré field to corresponding deformations as follows

$$U(x) = (1/2\pi f) \arccos [(I(x) - I_0)/ I_1] \tag{3}$$

where $I(x)$ is the light intensity at the point under consideration, I_1 is the intensity amplitude of the harmonic term in the optical law for the field and I_0 is the average background intensity. Equation (3) applies only over any one half fringe, i.e. over any one change from dark to bright or from bright to dark. The equation is useful in determining the displacements directly once the quantities $I(x)$, I_1 and I_0 are determined by the image analyzer for a certain half fringe. It should be noted that equation (3) is an ideal representation of the formation of moiré fringes. It does not assume any noise, drifts, or imperfections in in the pattern. The presence of optical noise in moiré patterns is often encountered. Noise could be due to image imperfections caused by dirt, scratches, fingerprints, or air bubbles in the grating replica. Noise may also be caused by low quality, scratched or damaged optics. The presence of optical noise slightly alters the appearance of light intensity distributions along some lines in a moiré field. Spikes or irregular changes in the distribution may be present. An example of a contaminated distribution and the fringe pattern from which it was read, are shown in Figure 3a,b. The case shown is an extreme case and is not often encountered. It is presented for demonstration purposes only. Usually the level of noise is much less, particularly when proper care is given to specimen handling and to optics cleanliness. As mentioned earlier, moiré fringes are constructed by one order of diffraction only and are, therefore, expected to be smooth continuous distributions of light intensity. This gives a general impression that the additional wiggles in a moiré pattern - such as those in Figure 3b - are not displacement related but rather noise related. Their elimination is, therefore, justified. A digital filter based on the Fast Fourier Transform (FFT) has been developed and used as a tool for noise elimination. The original contaminated distribution is represented by a Discrete Fourier Transform. The spectrum of the Fourier coefficients is examined, higher order coefficients are eliminated and the distribution is reconstructed using an inverse Fourier Transform based on the remaining coefficients only. Figure 3c, show the spectrum for the distribution given in Figure 3b. The decision on where to terminate the Fourier expansion is based on the quality of the specific image under consideration. The process is carried out incrementally. First a high number of Fourier coefficients is used to reconstruct the distribution which is examined for presence of noise. If noise is still there, the number of terms taken is reduced and the procedure is repeated if necessary until an optimum distribution combining minimum noise and no changes in shape, is reached. For the example shown, it was sufficient to retain only 10 coefficients of the 64 representing the unfiltered distribution. The reconstructed filtered distribution is then used for displacement computation. The use of digital filters in moiré interferometry and their effect on the displacement data has been investigated in detail [8]. They were found to be an efficient tool in managing noisy moiré fringes.

Fractional Fringe Moiré, has eliminated the need for high fringe density (closely spaced fringes) [9] usually required for accurate strain analysis and has encouraged the study of cases of low fringe density (small deformations) as in electronic devices. In addition, the use of digital image processing has enhanced the technique and increased the sensitivity by almost two orders of magnitude [10], while making the process of the fringe analysis more expedient and accurate with less human interference.

4. SPECIMEN AND TESTING RIG

The specimen was prepared from AT&T 1MB DRAM device by slicing the package along selected planes to expose the chip and the lead frame, Figure 4. Crossed gratings of a frequency of 1200 lines/mm in the horizontal and vertical directions were replicated on the specimens. Specimen was placed at room temperature in a custom built oven capable of heating the specimen uniformly from all directions. The oven was equipped with a variable power heating element to allow for different heating rates as well as a precision temperature controller to maintain steady state conditions, if desired.

The experimental setup used is schematically illustrated in Figure 5. Light from an Helium-Neon laser (Spectra Physics model 127-25 mW) is decollimated by a convex lens, cleaned from off-axis reflections through a spatial filter and collimated again to a large beam diameter (50 mm) using a parabolic mirror. The expanded beam is then projected on the specimen at the proper angle α which defines the frequency of the reference grating. The second beam needed to create the virtual grating is produced by a plane mirror that reflects part of the beam back to the specimen at an angle $-\alpha$. For simplicity the figure shows the optical arrangement for monitoring one displacement component only, to get the other component the specimen has to be rotated through a right angle or better yet a second identical optical arrangement sensitive to the other field may be superimposed on the basic one with the specimen remaining stationary in place. The latter arrangement; often referred to as "Four Beam Interferometer" [6] necessitates optical separation of the two fields by means of polarizing filters. The interference pattern emerging from the specimen is collected by a video camera that is connected to a PC-based digital image processor.

5. PROCEDURE

At room temperature the specimen is inserted in the oven and the apparatus is adjusted for proper alignment of all optical components and proper specimen orientation. Several optical effects are observed to assure that laser beams make the correct angles of projection that would yield a virtual grating frequency of 2400 lines/mm. Moreover, specimen orientation is further adjusted such that the lines of specimen grating are parallel to those of the virtual grating for the component of interest by rotating the specimen about an axis perpendicular to the plane under investigation. Also, the optical axis of the viewing camera is adjusted to coincide with the outward normal to the specimen to assure that recorded images are not inclined projections of actual patterns. If these adjustments are properly carried out, there will be no fringe pattern in the camera view. Practically, some deviations from ideal settings might exist or some of the optical components may have some irregularities. This usually yields an initial fringe pattern of relatively small number of fringes (null or zero field). If such a field exists, its effect has to be subtracted from the final displacement field to yield the net load induced displacements. For each specimen the null field (if any) is recorded for each of the displacement components (*U and V*), then the heaters are turned on and the specimen is allowed to heat quasi-statically (over about 20 minutes) until a final temperature of 90 °C is reached. There is no constraints on the final temperature other than the maximum temperature the epoxy used for grating replication may stand. There are several types of high temperature epoxies that may be used should the application require reaching high temperature levels. For the specimen under consideration the epoxy used (Devcon 2-Ton Epoxy) can safely stand temperatures as high as 120 °C. Based on that, a final temperature of 90 °C was selected. The final corresponding fringe pattern is also recorded for each component separately. The obtained fringe patterns are then filtered and analyzed using a specially written image processing software for fast and accurate investigation of light intensity distributions from which the corresponding thermally induced displacements are computed in areas of interest of the specimen. If a null field exists, its displacements are subtracted from the corresponding final field automatically by the same software. The specimen is then allowed to cool down back to room temperature and the resulting null field is visually compared to the one recorded prior to heating to make sure that the alignment of the system has not been disturbed during the course of the experiment and also to confirm that the replica has not failed by heating. Once displacement data are available the total strains are computed by differentiation of the displacement distributions with respect to the two basic directions; horizontal (x) and vertical (y). These strains are given by

$$\varepsilon_x = \frac{\partial U}{\partial x}, \quad \varepsilon_y = \frac{\partial V}{\partial y}, \quad \gamma_{xy} = \frac{\partial U}{\partial y} + \frac{\partial V}{\partial x} \tag{4}$$

The above equation gives the components of the total strain tensor which includes the free thermal expansion of the material. To get the net mechanical strain, i.e. strain due to mismatch in the coefficients of thermal expansion of the different materials in the package, the free expansion strain for each material ($\alpha \ \Delta T$) has to be subtracted from the total strain, thus giving the net mechanical strains as

$$\varepsilon_{x,m} = \varepsilon_x - \alpha \ \Delta T, \quad \varepsilon_{y,m} = \varepsilon_y - \alpha \ \Delta T, \tag{5}$$

where subscript m denotes mechanical strain only.

6. RESULTS

An example of the fringe patterns at 90 °C is shown in Figure 6, for the left half of the package. For this particular specimen the room temperature was 24.5 °C and the null field had no fringes for the horizontal direction. Thus, the pattern for the horizontal component shown in Figure 6.a, is the final contour map of axial displacements (U) due to heating. On the other hand, the null field for the vertical direction, Figure 6.b, had several fringes which have to be taken into consideration when processing the final fringe pattern for the vertical displacement component at 90 °C, Figure 6.c. Each fringe (dark line) is a line of constant displacement, i.e. all points along that line have undergone the same amount of displacement in the direction considered. To assign fringe values, a reference point of zero displacement need to be known. Generally this could be any fixed point of the structure or a boundary point of known displacement. In our case the device was not constrained at any point and was free to expand in all directions. Therefore, any arbitrary point can be considered as a reference point and all displacements can be measured relative to it. For analysis of strains the reference point displacement is immaterial, since only relative displacements are needed for strain computations, equation (4), and not the absolute values. The lower left corner of the package was considered the reference point in this example. Fringes passing through it were assigned the order zero and all other fringe orders were counted relative to that point with orders increasing in the positive directions of the reference axes. The assigned fringe orders are also shown in Figures 6a, 6b & 6c.

Strains were derived from deformation fields according to equation (4) and the resulting total horizontal strains ε_x are shown in Figure 7a. Upon subtraction of the free thermal expansion term the net mechanical strain $\varepsilon_{x,m}$ becomes as shown in Figure 7b. Similarly, normal strains in the vertical direction based on the net vertical deformation field (after subtracting the null field deformations) become as shown in Figure 8a & 8b. The total strain ε_y is shown in Figure 8a while the net mechanical strain is shown in Figure 8b. In addition the shear strains γ_{xy} have also been computed and the corresponding contour map is presented in Figure 9. There is no need to use subtraction as in equation (5) for the shear strains as long as all materials are thermally isotropic; which is the case under consideration.

Figures 7b, 8b and 9 give the full strain picture and show areas and levels of high and low strains. Clearly the silicon chip is under tension in the horizontal and the vertical directions with the highest strain level at the chip corner.

Various problems in electronic packages may be efficiently assessed by this technique since it renders realistic, accurate and valuable full field data that are necessary for any design, redesign, testing and evaluation processes of electronic packages. It is obvious that the technique possesses an excellent potential for serving as a basic deformation monitoring tool for the microelectronics industry. It should be noted that FFMI is independent of the type of loading applied to the specimen. Any type of loading whether thermal, mechanical or a combination of both may be applied as long as it induces deformations in the structure considered. The technique is sensitive to deformation only and has excellent applicability to numerous situations whether in electronic, military, civil, aerospace, mechanical or materials applications.

7. CONCLUSIONS

Full displacement and strain fields in electronic devices subjected to uniform heating were rendered measurable by means of FFMI. Components of the net mechanical strain tensor due to the thermal mismatch in the coefficients of thermal expansion of the different materials in an electronic package were accurately monitored from displacement measurements in the two basic directions of the package. The use of fractional fringe orders has eliminated the need for high fringe densities and improved the displacement sensitivity thus highly enhancing existing moiré interferometry.

FFMI being an accurate, sensitive and reliable technique known for full field displacement measurements proved to be a successful tool for strain analysis in electronic packages. It has the advantage of being a more realistic tool for deformation studies, over many analytical and numerical models for electronic components that imply many simplifying assumptions. From the resulting displacement fields, full strain and stress fields may be easily derived. This information is essential for package design, assessment of the different material combinations in a package, modifications for optimum package geometry and for better understanding and selectivity of the different materials used in an electronic package.

8. ACKNOWLEDGEMENT

This work has been supported in part by the Semiconductor Research Corporation, Research Triangle, NC. Their support is gratefully appreciated. The authors are also grateful to the Packaging Group of AT&T Bell Laboratories in Allentown, PA, for their generous support in specimen preparation and fruitful discussions regarding stress related problems in electronic packages.

9. REFERENCES

1. B. S. H. Royce, "Differential Thermal Expansion in Microelectronic Systems," IEEE Trans. Comp., Hybrids, Manuf. Technol., vol. 11, no. 4, pp 454 - 463, 1988.

2. S. Groothius, W. Schroen and M. Mortuza, "Computer Aided Stress Modelling for Optimizing Plastic Package Reliability," 23rd Annual Proceedings, Reliability Physics Symposium, 1985.

3. H. Kotake and S. Takasu, "Quantitative Measurements of Stress in Silicon by Photoelasticity and its Application," J. Electrochem. Soc.: Solid-State Science and Technology, pp. 179 - 184, 1980.

4. J. Spencer, "Calculating Stress and Mobility in Silicon Chips Using Strain Gauge Measurements'" Semiconductor Engineering Journal, vol. 1, pp 34-37, 1981.

5. D. Post and W. A. Baracat, "High-sensitivity Moiré Interferometry- A Simplified Approach," Experimental Mechanics, vol. 21, no. 3, pp. 100-104, 1981.

6. D. Post, " Moiré Interferometry at VPI & SU," Experimental Mechanics, vol. 23, no. 2, pp 203 - 210, 1983.

7. C. A. Sciammarella, "Basic Optical Law in the Interpretation of Moiré Patterns Applied to the Analysis of Strains-Part *I*," Experimental Mechanics, vol. 5, pp. 154-160, 1965.

8. A. F. Bastawros, "Enhancement of Moiré Interferometry by Digital Image Processing," Ph. D. Dissertation, Lehigh University, 1990.

9. A. S. Voloshin, C. P. Burger, R. E. Rowlands and T. S. Richard, "Fractional Moiré Strain Analysis using Digital Imaging Techniques," Experimental Mechanics, vol. 26, pp. 254-258, March 1986.

10. A. F. Bastawros, A. S. Voloshin and P. Rodogoveski, "Experimental Validation of Fractional Fringe Moiré Interferometry," Proc. 1989 SEM Spring Conf. on Experimental Mechanics, pp 401-406, Cambridge, MA, 1989.

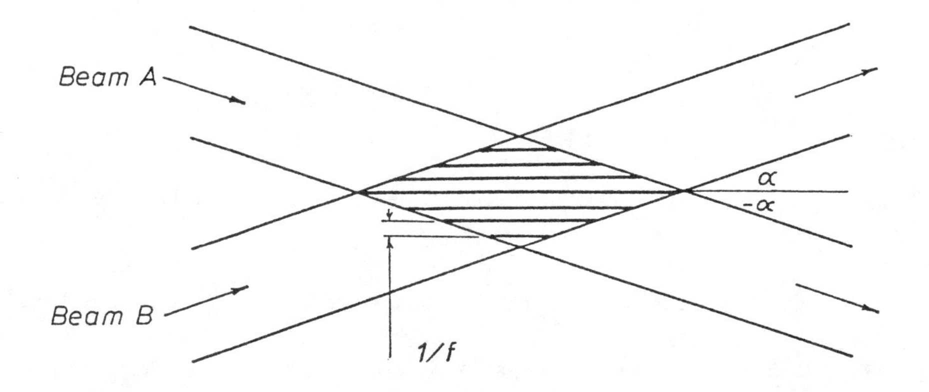

Fig. 1 Formation of virtual gratings by the intersection of two coherent laser beams.

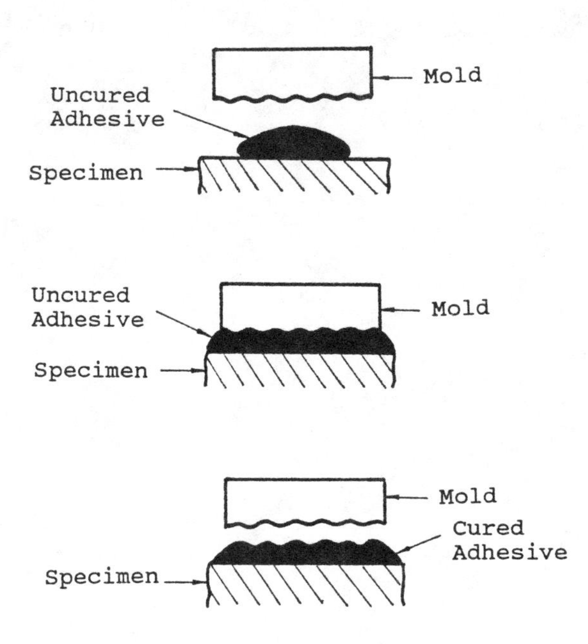

Fig. 2 Replication of specimen gratings form a mold.

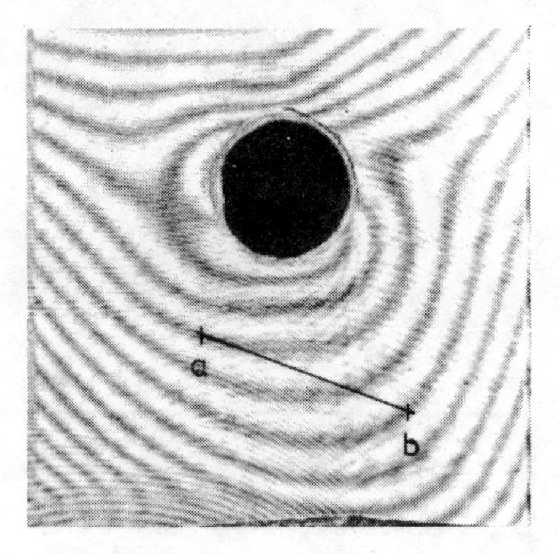

Fig. 3a Example of a low quality moiré pattern.

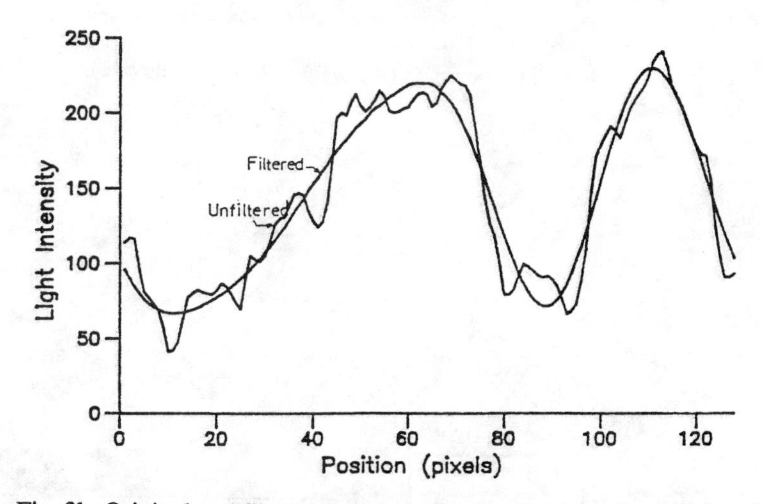

Fig. 3b Original and filtered light intensity distributions along line ***a - b***.

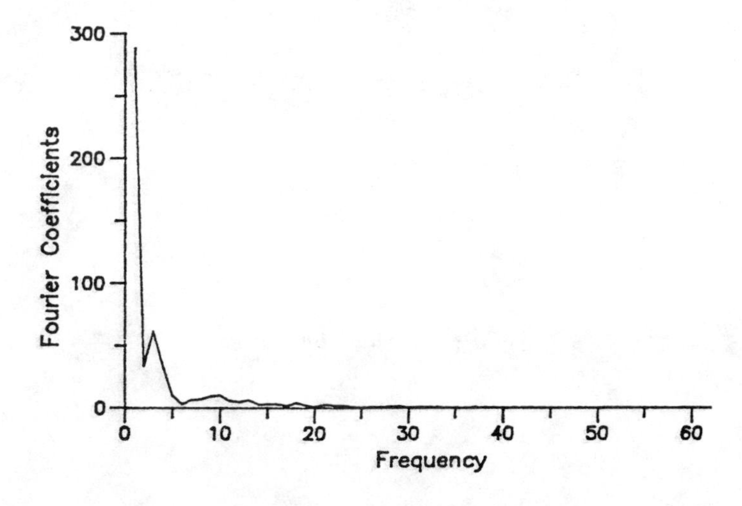

Fig. 3c Spectrum of Fourier coefficients for distribution along line ***a - b***.

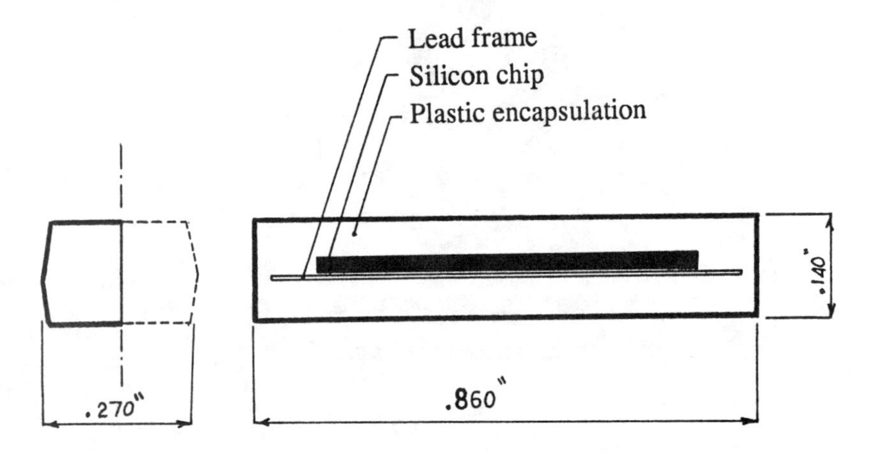

Fig. 4 Details of test specimen (AT&T 1MB DRAM device).

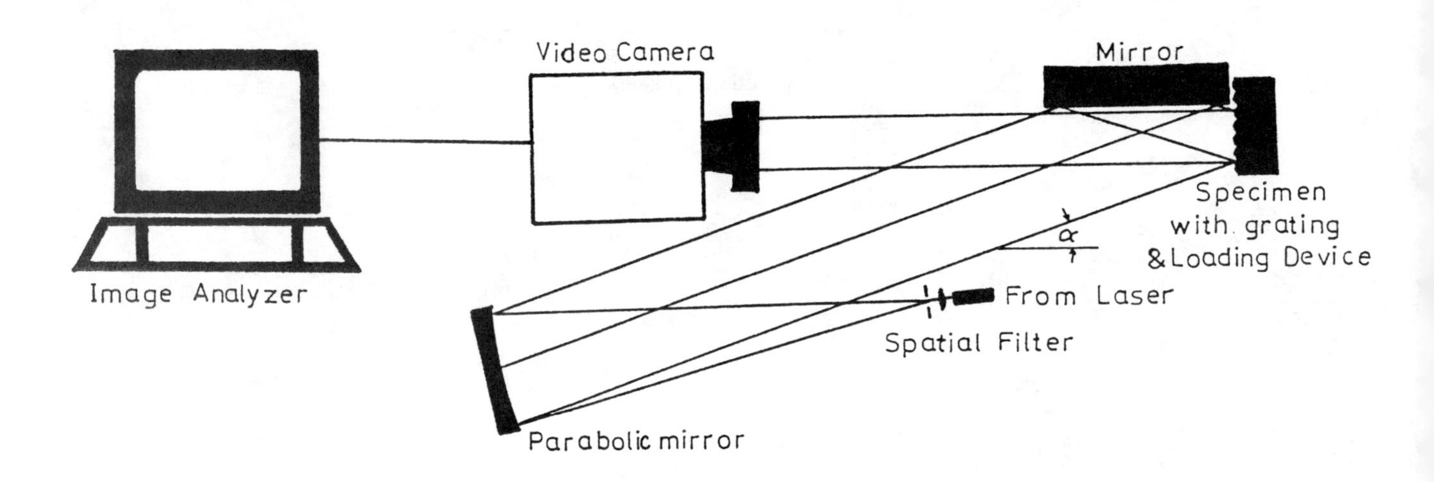

Fig. 5 Optical system for Moiré Interferometry.

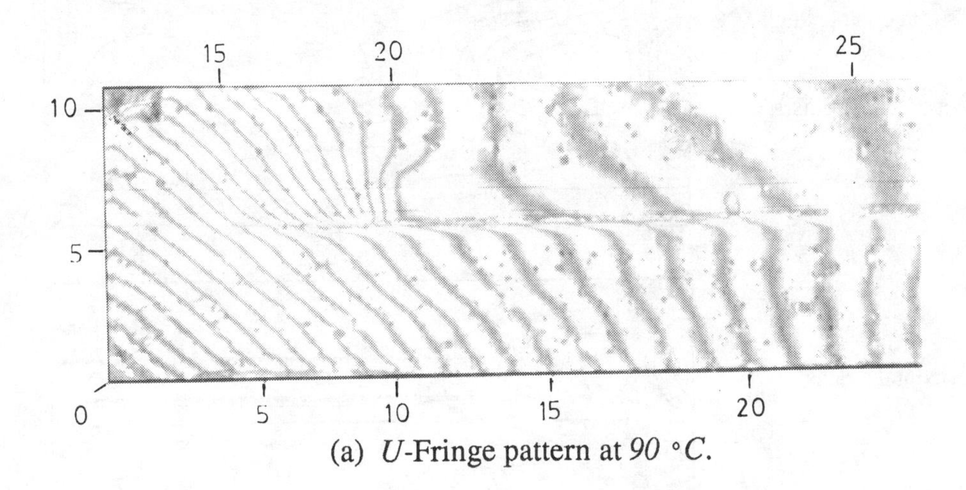

(a) *U*-Fringe pattern at *90* °*C*.

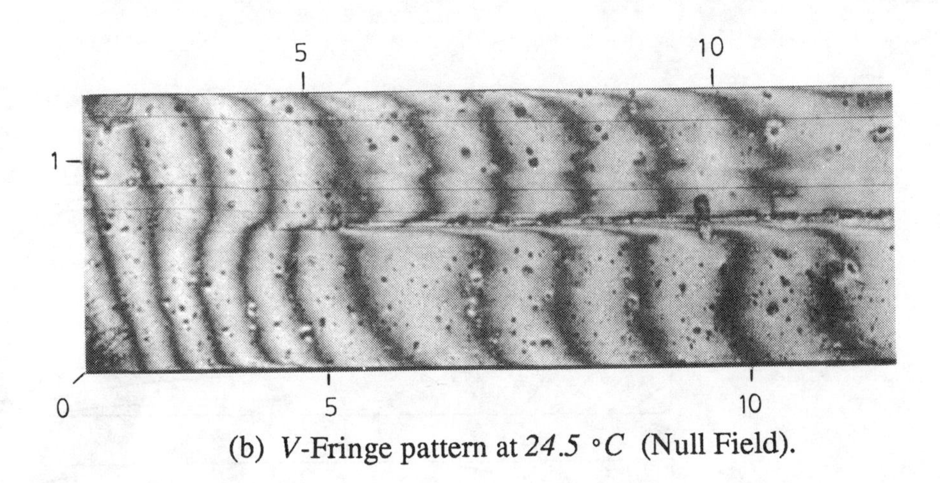

(b) *V*-Fringe pattern at *24.5* °*C* (Null Field).

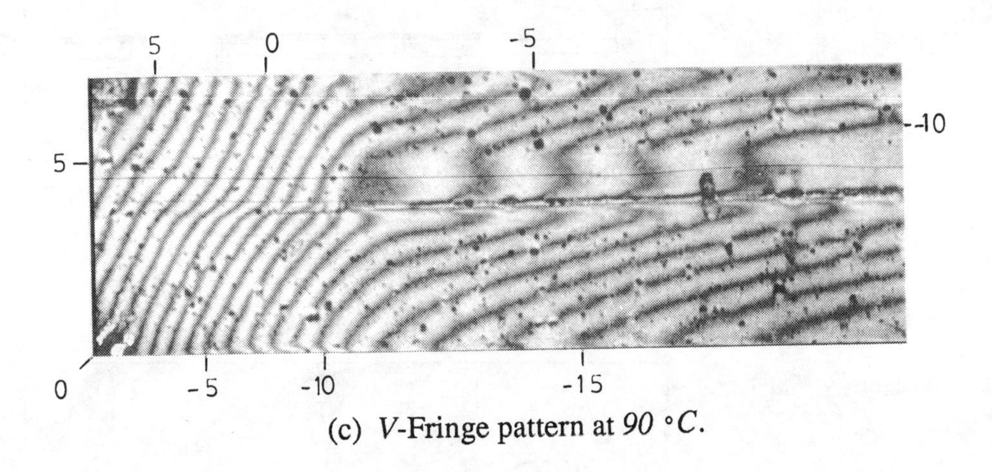

(c) *V*-Fringe pattern at *90* °*C*.

Fig. 6 Typical fringe patterns in a 1MB-DRAM specimen.

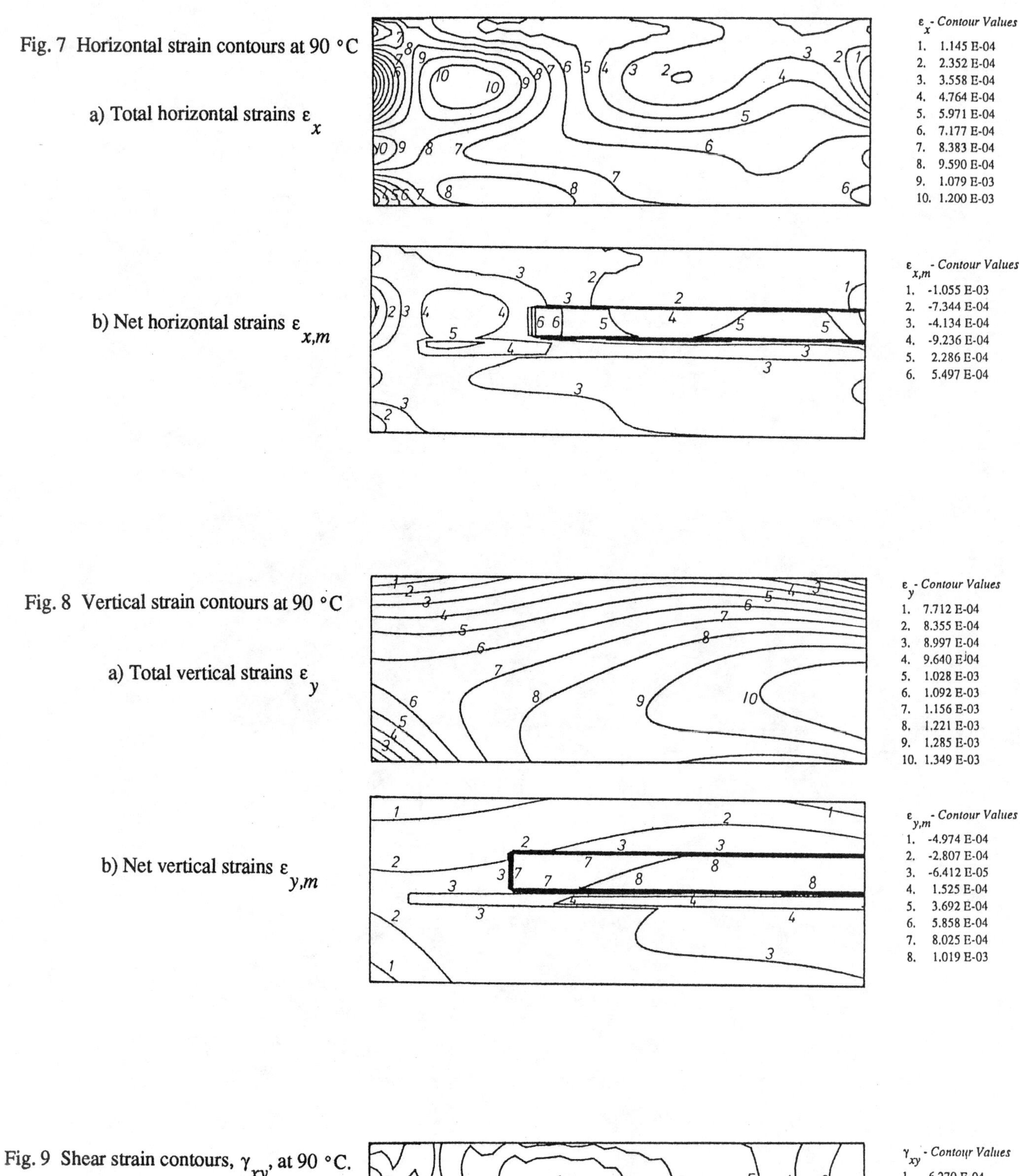

Fig. 7 Horizontal strain contours at 90 °C

a) Total horizontal strains ε_x

b) Net horizontal strains $\varepsilon_{x,m}$

Fig. 8 Vertical strain contours at 90 °C

a) Total vertical strains ε_y

b) Net vertical strains $\varepsilon_{y,m}$

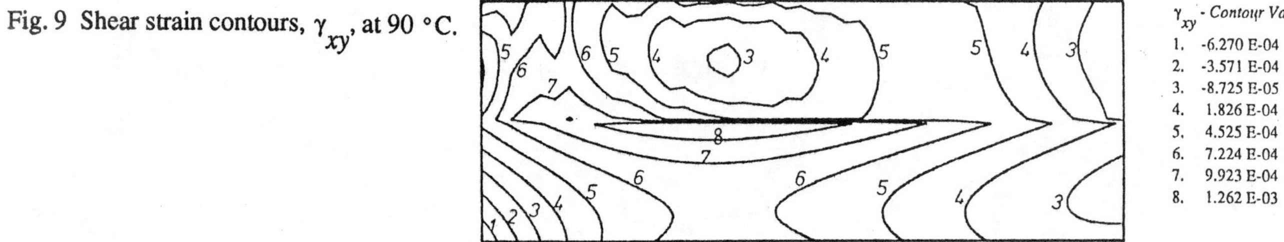

Fig. 9 Shear strain contours, γ_{xy}, at 90 °C.

Absolute Interferometric Testing of Spherical Surfaces

Bruce E. Truax, Zygo Corporation
Laurel Brook Road, Middlefield, CT 06455

ABSTRACT

In typical interferometric testing the part under test is measured against a reference standard. The measured result is the difference between the errors in the test and reference surfaces plus any additional errors introduced by the interferometer. For accurate qualification of the reference surface it is necessary to employ a technique that can measure the part absolutely. This paper examines an existing technique[1] for absolute testing of spherical surfaces which produces a map of the entire surface. The capabilities of this technique, error sources, and experimental data will be examined.

1. INTRODUCTION

It is well known that interferometry of optical surface figure is a relative measurement. An interferometer measures the difference in surface shape between the surface under test and, in the case, of a Fizeau Interferometer, the reference surface. For other interferometer configurations such as a Twyman-Green, the errors of additional optics in the cavity are also included in the measurement. For this reason the most common, commercially available interferometers are of the Fizeau configuration and they are used with high quality reference optics. Typical reference optic quality is 1/20th of a wavelength at 0.6328 μm yielding a measurement accuracy of about 1/10th wavelength. Obtaining higher accuracy measurements requires absolute surface measurement techniques. These absolute techniques are used either by the metrologist to test the surface in question or by the fabricator of the reference optic to produce and guarantee higher accuracy reference surfaces. There are two techniques available for performing absolute surface measurement. One technique is known as the 3-Flat method and is used to absolutely measure a diameter on a flat surface. The second technique, and the subject of this paper, is known as spherical surface certification and it is used to absolutely measure the entire surface of a spherical optical element.

The technique of spherical surface certification was first described by A. F. Jensen in 1973 for Twyman-Green interferometers.[2] It was later referenced in two papers. There is no published reference with the derivation of the formulae.[3,4] This paper presents the derivation of the formulae used in this procedure as applied to a Fizeau interferometer as well as some actual test results using this technique with a commercially available interferometer.

2. THEORY

A greatly simplified schematic of a Fizeau interferometer is shown in Fig. 1.

The incoming wavefront of coherent light is assumed to have no aberration and is represented by $W_I(r,\theta)e^{i(kz+\omega t)}$. This beam passes through beam forming optics, the focusing lens, the reference surface and finally reflects from the test surface. Each of these surfaces adds aberration to the wavefront. It is assumed that the input beam and all apertures are circular and that the phase aberrations can be represented by circular polynomials of the form.

$$\vartheta(r,\theta) = \sum_{n=0}^{\infty} \sum_{m=0}^{\infty} C_{nm} r^n (\cos m\theta + \sin m\theta) \qquad (1)$$

where r is normalized to 1 at the edge of the aperture.

---

[1] Jensen, A.E., "Absolute Calibration Method for Laser Twyman-Green Wavefront Testing Interferometers," Paper ThG19, Fall OSA Meeting, Rochester, N.Y (October 1973).
[2] Ibid.
[3] Bruning, J. H., et al., "Digital Wavefront measuring Interferometer for Testing Optical Surfaces and Lenses," Applied Optics, Vol. 13 No. 11, pp. 2693-2703 (Nov. 1974) .
[4] Berman, J., Hunter, G.C., Truax, B.E., "Interferometric Lens Testing to λ/40 at λ=0.442 Microns," Presented at Optical Fabrication and Testing Conference, Cherry Hill, New Jersey (June 1985).

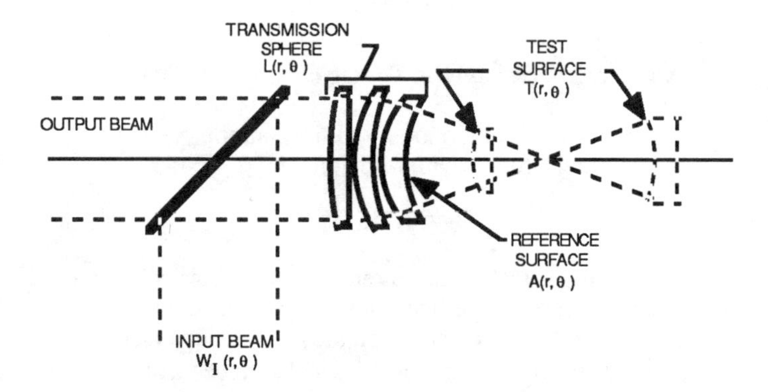

Figure 1. Interferometer Schematic

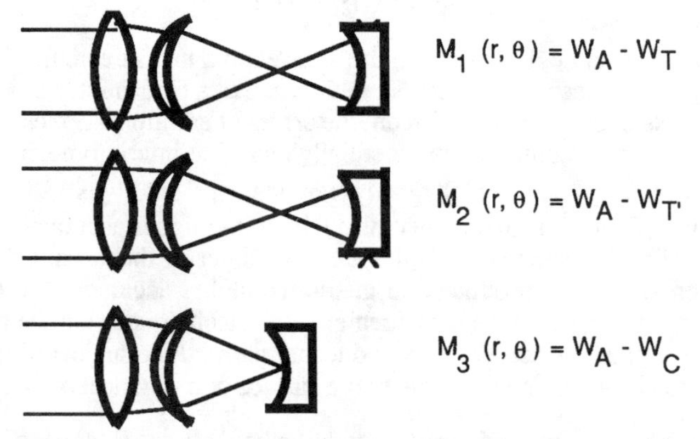

where $W_A(r,\theta)$ is the wavefront reflected from the aplanat,
$W_T(r,\theta)$ is the wavefront reflected from the test surface,
$W_{T'}(r,\theta)$ is the wavefront reflected from the test surface after 180° rotation,
$W_C(r,\theta)$ is the wavefront reflected from the cat's eye reflection.

Figure 2. Spherical Certification Measurements

The absolute measurement of the test surface $T(r,\theta)$ requires three measurements as shown in Fig. 2.

Using Figs. 1 and 2, Eq. (1), and the fact that the aberrations introduced by each element in the optical path add linearly, the results of the measurements can be calculated. First calculate each wavefront component.

$$W_A(r,\theta) = I(r,\theta) + 2\,L(r,\theta) + 2n\,A(r,\theta)\,. \tag{2}$$

$$W_T(r,\theta) = I(r,\theta) + 2\,L(r,\theta) + 2(n-1)\,A(r,\theta) + 2\,T(r,\theta)\,. \tag{3}$$

$$W'_T(r,\theta) = I(r,\theta) + 2\,L(r,\theta) + 2(n-1)\,A(r,\theta) + 2\,T(r,\theta + 180^\circ)\,. \tag{4}$$

$$W_C(r,\theta) = I(r,\theta) + L(r,\theta) + (n-1)\,A(r,\theta) + (n-1)\,A(r,\theta + 180^\circ) + L(r,\theta + 180^\circ)\,. \tag{5}$$

Where $I(r,\theta)$ is the wavefront error of the source,
$L(r,\theta)$ is the wavefront error introduced by the focusing optics in the aplanat,
$A(r,\theta)$ is the error in the aplanat reference surface, and
$T(r,\theta)$ is the error in the surface under test.

Since $\cos m\theta = \cos m(\theta + 180°)$ and $\sin m\theta = -\sin m(\theta + 180°)$ for m odd, Eq. (5) can be reduced to

$$W_C(r,\theta) = I(r,\theta) + \begin{cases} 2\,L(r,\theta) + 2\,(n-1)\,A(r,\theta) & m \text{ even} \\ 0 & m \text{ odd} \end{cases} \quad (6)$$

Next, combine the wavefronts into the individual measurements.

$$M_1(r,\theta) = W_A(r,\theta) - W_T(r,\theta). \quad (7)$$

$$M_1(r,\theta) = \Big[\, I(r,\theta) + 2\,L(r,\theta) + 2n\,A(r,\theta) \,\Big] - \Big[\, I(r,\theta) + 2\,L(r,\theta) + 2\,(n-1)\,A(r,\theta) + 2\,T(r,\theta)\Big]. \quad (8)$$

$$M_1(r,\theta) = 2\,A(r,\theta) - 2T(r,\theta). \quad (9)$$

$$M_2(r,\theta) = W_A(r,\theta) - W'_T(r,\theta). \quad (10)$$

$$M_2(r,\theta) = \Big[\, I(r,\theta) + 2\,L(r,\theta) + 2n\,A(r,\theta) \,\Big] - \Big[\, I(r,\theta) + 2\,L(r,\theta) + 2\,(n-1)\,A(r,\theta) + 2\,T(r,\theta + 180°)\Big]. \quad (11)$$

$$M_1(r,\theta) = 2\,A(r,\theta) - 2T(r,\theta + 180°). \quad (12)$$

$$M_3(r,\theta) = W_A(r,\theta) - W_C(r,\theta). \quad (13)$$

$$M_3(r,\theta) = [\, I(r,\theta) + 2\,L(r,\theta) + 2n\,A(r,\theta)\,] - I(r,\theta) + \begin{cases} 2\,L(r,\theta) + 2\,(n-1)\,A(r,\theta) & m \text{ even} \\ 0 & m \text{ odd} \end{cases} \quad (14)$$

$$M_3(r,\theta) = \begin{cases} 2\,A(r,\theta) & m \text{ even} \\ 2\,L(r,\theta) + 2n\,A(r,\theta) & m \text{ odd} \end{cases} \quad (15)$$

Now that the three measurements have been made, two intermediate computations are performed. In the first computation $M_3(r,\theta)$ is added to itself after a rotation of 180°.

$$\Psi(r,\theta) = M_3(r,\theta) + M_3(r,\theta + 180°) \quad (16)$$

$$\Psi(r,\theta) = \begin{cases} 2\,A(r,\theta) + 2\,A(r,\theta + 180°) & m \text{ even} \\ 2\,L(r,\theta) + 2n\,A(r,\theta) + 2\,L(r,\theta + 180°) + 2n\,A(r,\theta + 180°) & m \text{ odd} \end{cases} \quad (17)$$

$$\Psi(r,\theta) = \begin{cases} 4\,A(r,\theta) & m \text{ even} \\ 0 & m \text{ odd} \end{cases} \quad (18)$$

The second intermediate computation is the sum of $M_1(r,\theta)$ and $M_2(r,\theta)$ after rotating $M_2(r,\theta)$ by 180°.

$$\Psi'(r,\theta) = M_1(r,\theta) + M_2(r,\theta + 180°). \quad (19)$$

$$\Psi'(r,\theta) = \Big[2\,A(r,\theta) - 2\,T(r,\theta)\Big] + \Big[\,2\,A(r,\theta + 180°) - 2\,T(r,\theta + 360°)\Big]. \quad (20)$$

$$\Psi'(r,\theta) = -4\,T(r,\theta) + \begin{cases} 4\,A(r,\theta) & m \text{ even} \\ 0 & m \text{ odd} \end{cases} \quad (21)$$

The solution is now obvious, by taking 1/4 of the difference between Ψ and Ψ', $T(r, \theta)$ is found.

To summarize, by referring to Fig. 2 and Eqs. (16) and (19), the following formula is derived.

$$T(r,\theta) = \frac{1}{4}\Big\{\Big[M_3(r,\theta) + M_3(r,\theta + 180°)\Big] - \Big[\,M_1(r,\theta) + M_2(r,\theta + 180°)\Big]\Big\} \quad (22)$$

This equation is easily implemented on phase measuring interferometers where the data is defined on a uniform grid.

3. EXPERIMENTAL RESULTS

Spherical surface certification has been implemented using a Zygo MARK IV interferometer, special software, and a precision mount. Software to implement Eq. (22) has been written using the MARK IV's programming capability. This software guides the user through the three measurements and makes provision for locating the optical axis. The precision mount, shown in Fig. 3, allows for translation of the part between the "cat's eye" and confocal positions as well as for the necessary 180° rotation of the test optic. In addition, a special reticle is provided that when used in conjunction with the software allows for accurate location of the optical axis.

Figure 4 illustrates an example of the results of this test procedure for testing an f/2 spherical surface. This same f/2 surface was measured using seven different transmission optics. A summary of the seven measurements is shown in Table 1. Note the excellent repeatability of data. The peak-to-valley errors have a spread of 0.02λ.

Meas. #	PV	rms
1	.057	.007
2	.064	.008
3	.063	.009
4	.044	.006
5	.046	.007
6	.054	.008
7	.053	.007
Std. Dev.	.008	.001

Table 1. Experimental Repeatability

This same technique has been used by the author's company to successfully fabricate and certify f/1 spherical surfaces to 0.025λ at λ=442nm.[1]

Figure 3. Photograph of Precision Mount for Spherical Surface Certification

[1] Berman, J., Hunter G.C., Truax, B.E., "Interferometric Lens Testing to λ/40 at λ=0.442 Microns'" Presented at Optical Fabrication and Testing Conference, Cherry Hill, New Jersey (June 1985).

*** CONFOCAL ***

03-AUG-1988/12: 22: 47
Part ID : CONFOCAL
Serial # : CONFOCAL
Fast : ON
Averages : 4
Trim : 0
Calibrate : ON
AGC : ON
Scale : 0.50
Wave Out : 0.6328
Remove : TLT PWR

PV : 0.106 PTS : 22217
RMS : 0.009

POWER : 0.0030

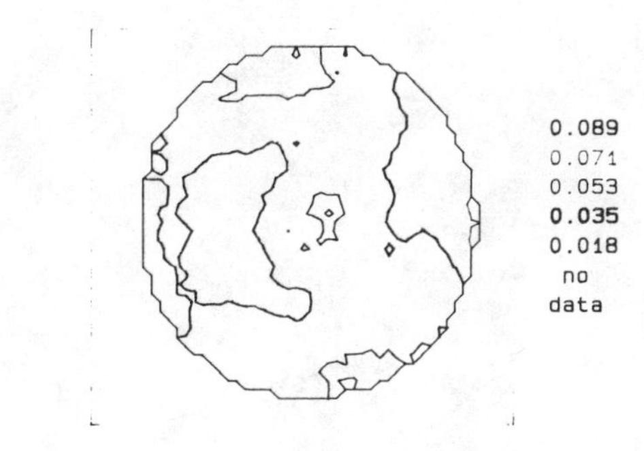

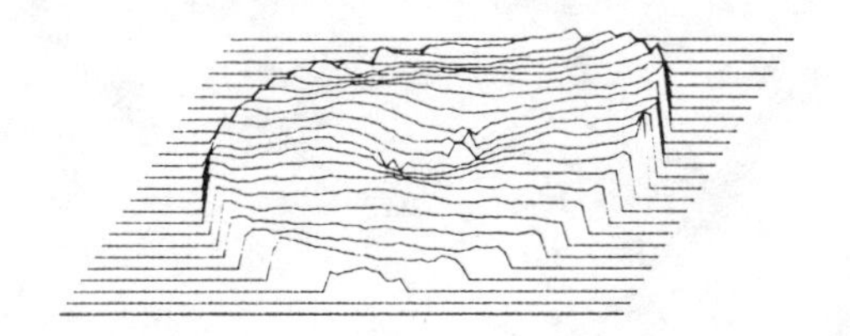

*** CONFOCAL (180) ***

03-AUG-1988/12: 31: 50
Part ID : CONFOCAL (180)
Serial # : CONFOCAL (180)
Fast : ON
Averages : 4
Trim : 0
Calibrate : ON
AGC : ON
Scale : 0.50
Wave Out : 0.6328
Remove : TLT PWR

PV : 0.123 PTS : 22258
RMS : 0.008

POWER : -0.0091

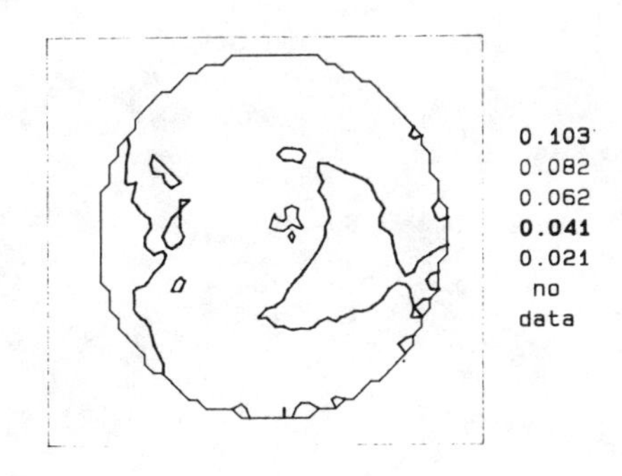

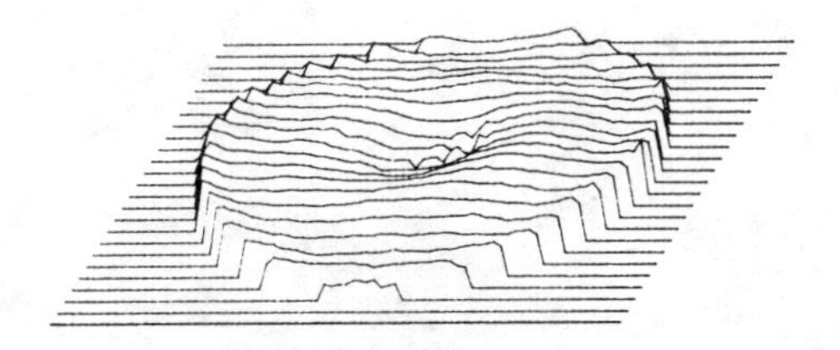

Figure 4. Experimental Results

*** CAT'S EYE ***

03-AUG-1988/12: 19: 37
Part ID : CAT'S EYE
Serial # : CAT'S EYE
Fast : ON
Averages : 4
Trim : 0
Calibrate : ON
AGC : ON
Scale : 0.50
Wave Out : 0.6328
Remove : TLT PWR

PV : 0.096 **PTS : 30715**
RMS : 0.014

POWER : -0.0810

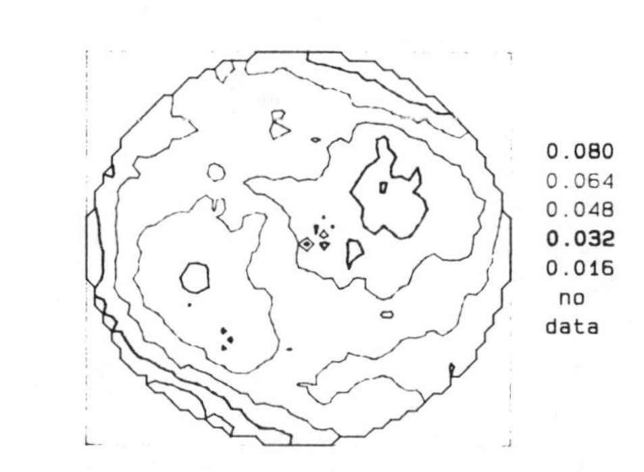

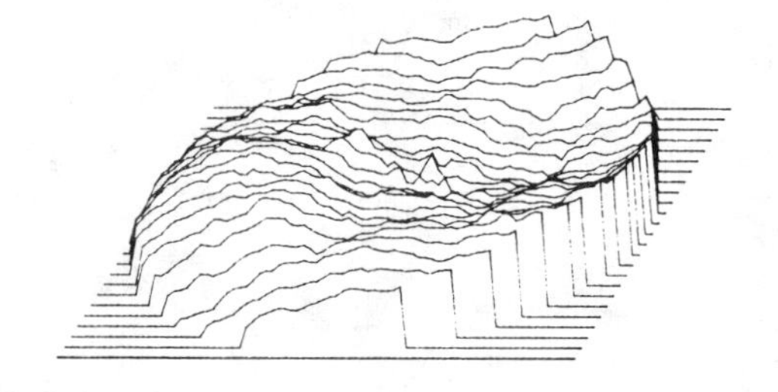

*** TEST PART ***

03-AUG-1988/12: 19: 37
Part ID : f/2.0 TS
Serial # :
Fast : ON
Averages : 4
Trim : 0
Calibrate : ON
AGC : ON
Scale : 0.50
Wave Out : 0.6328
Reference : DEMO LAB f/.75 TS
Remove : TLT PWR

PV : 0.060 **PTS : 21707**
RMS : 0.008

POWER : 0.0390

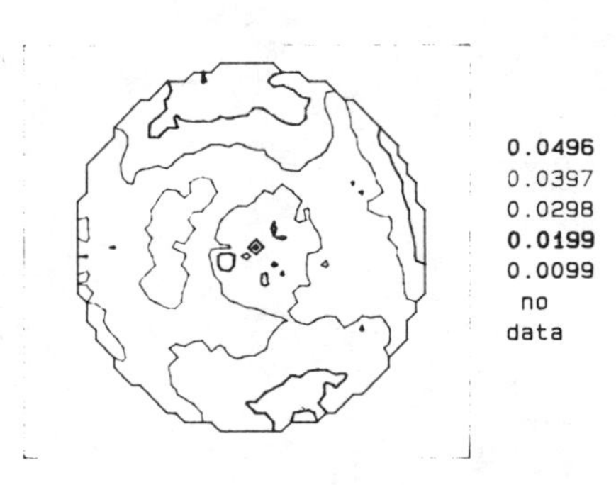

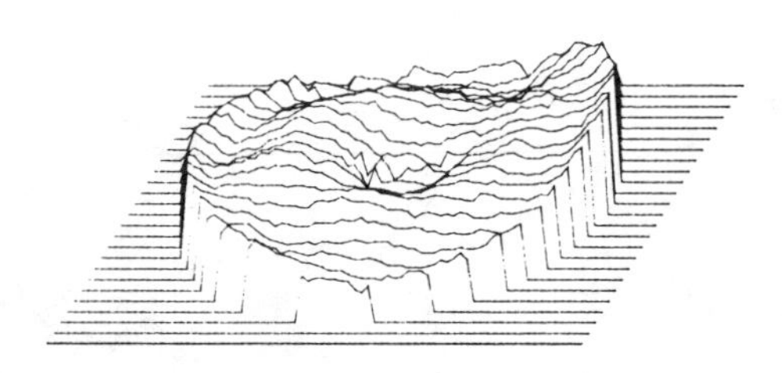

Figure 4. Experimental Results (Cont.)

4. SOURCES OF ERROR

4.1 Instrument Precision

As with any interferometric metrology vibration, thermal instability and electronic noise will cause random errors in the individual measurements that are then combined using Eq. (22) to yield the final measurement result. These errors can be minimized using proper vibration isolation and data averaging. The Zygo MARK IV instrument precision is better than 0.002λ rms for individual measurements, in a good environment. Equation (22) combines four measurements three of which are statistically independent. If the rotated "cat's eye" measurement is also considered independent for the purpose of this computation, then the resulting rms error due to these noise sources is less than 0.004λ rms.

4.2 Systematic Error

There are two sources of systematic errors that must also be considered, these are improper location of the optical axis and errors due to distortion when measuring with non-nulled interferometer cavities.

4.2.1 Optical Axis Location

Due to the data rotation that is necessary to compute the surface figure $T(r,\theta)$, proper location of the optical axis is critical if accurate results are to be obtained. Incorrect location of the optical axis causes a wavefront shear which introduces errors that are proportional to the errors present in the individual wavefronts. When testing high quality optics with a Fizeau interferometer, the measurement with the most aberration is the "cat's eye" measurement. Typically this measurement will have significant coma (all aberrations proportional to even multiples of θ cancel, see Eq. (15), except for errors in the reference surface which are usually quite small). The coma should cancel as shown in Eq. (18) when M_3 is added to itself after 180° rotation of the data. If the optical axis is located incorrectly, astigmatism will be introduced into the final result due to the lateral shear of the comatic wavefront. The magnitude of this error can be expressed as

$$\varepsilon = 3\delta C \cos\varphi \tag{23}$$

where
δ=optical axis location error expressed as a fraction of the aperture diameter
C=magnitude of 3rd order coma
φ=angle between the coma and the line formed by the table optical axis and the chosen optical axis.

As a simple example, if $\varphi=0$, $C=0.5\lambda$ and $\delta=0.005$ then $\varepsilon=0.008\lambda$ of cylinder in the final result. This is beginning to be significant when measuring surfaces with accuracies of 0.025λ.

4.2.2 Non-Nulled Interference Cavities

Deriving the optimum accuracy from any phase measuring interferometer system can only be accomplished by making measurements with nulled interference cavities. Distortion present in the interferometer imaging system will introduce errors that are proportional to the slope errors of the wavefront being measured. When testing high quality surfaces these slope errors will be minimal as long as the tests are performed with the interference cavity nulled. As an example, take the case of simple defocus when combined with third order distortion. The wavefront form for defocus is shown in Eq. (24).

$$W(r,\theta) = k\, r^2 \tag{24}$$

where k is the defocus in waves of error at the edge of the aperture.

Third order distortion introduces an error into the position coordinate r of the form

$$r = r' + \zeta\, r'^3 \tag{25}$$

where ζ is the fractional distortion at the edge of the aperture.

Substituting Eq. (25) into Eq. (24) the form of the wavefront at the detector can be derived as in Eq. (26)

$$W'(r,\theta) = k\,[\, r^2 + 2\zeta r^4 + \zeta^2 r^6]. \tag{26}$$

The equation for the error that will be measured is simply the difference of Eqs. (24) and (26). Assuming that z is small the measured error can be expressed as

$$E(r,\theta) = 2k\zeta r^4. \tag{27}$$

For a simple case of one wave of defocus and 0.5% distortion the measurement error at the edge of the aperture (r=1) will be 0.01 wave. It is quite obvious that if the four measurements were each defocused by one wave in the same direction the error becomes significant. The distortion error of 0.5% is quite reasonable for high f-number optics, but as the f-number approaches f/1, or worse yet f/0.65, significant distortion is introduced due to the fact that the output angle of the transmission sphere is not linear with height. This nonlinearity introduces wavefront error at the rate of 0.17 waves for every wave of defocus when using an f/0.75 transmission sphere.

CONCLUSION

The absolute spherical certification procedure originally developed by Jensen for use with a Twyman-Green interferometer has been adapted for use with a Fizeau interferometer. Experimental results when testing an f/2 surface against a number of different transmission spheres demonstrates repeatability of better than 1/50 wave peak-to-valley and 1/300 wave rms. Various error sources are discussed indicating, as with any interferometric testing, that experimental technique is very important for obtaining accurate results.

ACKNOWLEDGMENTS

I would like to thank Peter Pfluke for performing the repeatability studies, Jill Duff for writing and testing the data reduction software, and Jim Soobitsky for the mechanical design of the precision mount.

Investigation of fiber reinforced plastics based components by means of holographic interferometry

Werner Jüptner, Thomas Bischof

BIAS, Klagenfurther Str. 2, 2800 Bremen 33, Germany

Abstract

Fiber reinforced plastic (FRP) is a modern material for light weighted constructions. However, by the inhomogenity of the material it is difficult to test it with conventional methods. This is valid as well for construction optimization as for non-destructive testing in production control. A good tool for the different tests is the holographic interferometry, because it invisions whole a field of deformations of a component under load.

The use of holographic interferometry is shown in two examples which cover as well construction optimization as non-destructive testing:

- Components of a radioastronomical antenna were investigated for construction optimization. The holographic interferometry was able to demonstrate the influence even of thin internal layers.
- An application with different aspects is the test of an aircraft component. In order to detect defects, special loadings and special equipment has to be developed.

The examples cover different aspects of testing fiber reinforced plastic components with holographic interferometry.

1. Introduction

Carbon-fiber reinforced plastic (CFRP) is a modern material which combines high strength with a low specific weight. Therefore components out of CFRP may be light weighted but strong and stiff, a quality especially important for large structures, e.g. in aircraft components and for the construction of a large antenna.

One disadvantage of the material is the problem of nondestructive testing because of the internal structure. Even if the component, in the example an aircraft aileron, can be tested during production, there are severe difficulties in the periodical tests. Holographic interferometry allows the component to be tested without dismounting any part.

In the example of the antenna panel, the task was the construction optimization of components of a huge radio astronomical antenna. Some of the deformation measurements, e.g. the quantification of the gravity influence, are nearly impossible with other methods. By means of holographic interferometry the stiffness of the panel could be improved up to a value sufficient to highest demands.

2. Holgraphic Interferometry

In holografic interferometry, the main tool is the holography. In order to take a hologram of an object, the laser beam is devided into two parts, fig. 1. One part is the object beam, which lightens the object. The light is back-scattered to a photographic plate called the hologram plate. The second part is lead directly to the hologram plate without any information. The two beams interfere and form a microscopic fringe system which carries all the information about the object. After the developement of the plate the hologram is replaced into the former position, fig. 2. By the illumination with only the reference wave the object is reconstructed with all details down to the surface structure. Now the same procedure can be done twice with the same object, fig. 3. When the object is changed before the second shot we can reconstruct the same object but slidely different. Then the two stages of the object will interfere with each other as shown by Stetson and Powell /1/ and the interferogram carries information about the change of the object before taking the second hologram, fig.4. The next slide shows an interference pattern in which the diffraction index of one of the light ways was changed, fig. 5. Such a interference carries a lot of information which can be evaluated only by means of a computer for quantitative evaluation or by vision control in nondestructive testing.

The holografic interferometry enables the evaluation of the dislocation vector field d(x,y,z) for the points (x,y,z) of a rough surface, as mentioned before. The object will be holographed twice once before and once during the application of the load. Both holograms - usually stored in the same photo-sensitive plate - will be reconstructed together, so that they interfere. During the two reconstructed stages of the object the surface is deformed. This leads to the change Δ of the light path from the light source over the object point (x,y,z) to the observation point given by, fig. 6:

$$\Delta(x,y,z) = \vec{d}(x,y,z) * [(\vec{b}(x,y,z)-\vec{s}(x,y,z)] \qquad (2.1)$$

$\vec{b}(x,y,z)$ and $\vec{s}(x,y,z)$ are the unity vectors in the illumination and in the observation direction. The intensity distribution in the interference pattern can be expressed as:

$$I(x,y,z) = a(x,y,z) + b(x,y,z) * \cos(\Theta(x,y,z)) \qquad (2.2)$$

with $\Theta(x,y,z) = 2\pi/\lambda * \Delta(x,y,z)$
λ : wave length of the laser

The main task in the evaluation of a holografic interference pattern is the estimation of the phase distribution Θ from the measured intensity distribution I and the pointwise solution of the equation system for the unknown:

$$\lambda/2\pi\ \Theta = \vec{d} * (\vec{b} - \vec{s}) \qquad (2.3)$$

For this task different methods are known /2,3,4/.

3. Optimization of Antenna Panels

In order to explore the very far universe, large radioastronomical antennas are build up. One of these radio telescopes is mounted by the IRAM, Institute for Radio Astronomic Measurements, France. Because the task was to reach objects in a distance of more than 10 billion light years, the quality and stability demands were high /5/. So, the contour should stay within a accuracy of better than 50 um under all loads, e.g. wind, temperature and gravity. To meet this requirements the whole antenna with a diameter of 30 m was constructed of smaller panels, size up to 1 m^2, fig. 7, which could be adjusted by piezo driven elements. But each of this smaller panels itself had to be stiff enough for the task. So different constructions of these panels were build up and had to be tested. The only tool powerful enough, was the holografic interferometry.

A special interferometer was build up, fig. 8. With this interferometer it was possible to turn whole the set-up 90° for the gravity measurement. The fringe system, fig. 9, was evaluated by measuring the fringe order/location data by hand. Afterwards these pairs of data were given into the computer. This is a simple method but it gave the necessary results, fig. 10. Preferable would be an automated fringe detection system as described later.

4. Flaw Detection on Aircraft Components

4.1 Laboratory Tests

The components provided for the holographic testing in the laboratory were prepared with defined flaws inside the structure. The main component was a so-called "Shell" with the dimensions 225*85 cm^2, fig. 11. It consists of a CFRP skin field reinforced with I-shaped stringers. The philosophy for the flaw preparation was to choose only ones which occur under normal production and operating conditions. There were amongst others: Defects in the bonds between the stringer and the skin field, fig. 12a, delaminations in different depths, fig. 12b, and separations of the upper stringer belt, fig. 12c. The tests in the laboratory should demonstrate the indications of the flaws in the interference patterns. For comparison served the interferogram of a flaw-free shell under thermal strain, induced by a direct current flow through the material, fig. 13. The change of temperature for the test is only 0,5 K. As expected, a periodic deformation and interference structure appears with smaller deformations above the stringers and larger ones in the areas between the stringers.

The structure of the interferograms changes drastically, when a shell with an artificial stringer separation is tested, fig. 14. A large ring system appears in the upper left of the interferogram corresponding to a 20 cm long stringer separation dislocated under the center of the ring system. The deformation caused by the flaw is about five times larger than the deformations in the surrounding areas. The area of this flaw reaches into that of the next stringer and overlaps with a second flaw area caused by a 10 cm long stringer separation there. The quantitative evaluation of the deformation perpendicular to the surface was evaluated by the phase-shift method with the BIAS image analysis system. The 3-D-plot of the deformation allows to distinguish the two different flaws, fig. 15.

Delaminations were simulated by introducing a sheet of teflon between the neighboured CFRP layers: the teflon disturbes the adhesion in the CFRP structure. Again, the test was performed by thermal strains. The internal delaminations were visible in several small areas, different to the stringer separation, which caused larger areas, fig. 16. The fault area is a complex system of overlapping distortion zones caused by different degrees of adhesion in the material. The distortion becomes less the deeper the flaw is located in the material: The resulting stress in the material is partially compensated or better distributed by the material above the defect. In some cases it was possible to detect flaws with a lateral length of 3 mm, but in general it was possible to detect delaminations down to a lateral length of 10 mm under one or two original CFRP layers, called "prepregs".

4.2 Holographic Measurements at a Airbus Aileron

In order to prove the practicability of the results gathered during the laboratory tests, holographic-interferometric measurements were performed under industrial conditions on an Airbus aileron during the periodic maintainance of one of these units. This aileron is, with its length of about 10 m, currently the largest standard CFRP component in the production of aircrafts.

In order to enable on-site measurements on such a large test object, a nonvibration-isolated holographic equipment had to be built-up, based on the phase reference mirror method /6/. The test object was loaded with thermal strains by heating it from inside with a hot air blower. The analysis of the interference pattern was performed quantitatively by the phase shift method with the BIAS image-processing system. First, the normal deformations, i.e. dislocation perpendicular to the surface, were evaluated at a temperature difference of 1,0 K, fig. 17. The false-color representation demonstrate the periodic structure of the component with minima of deformation above internal reinforcements by stringers. In this case, the component is flaw-free. Therefore no inhomogenities in the interferogramm are detectable.

5. Remarks on the Deformation Mechanism

In order to understand and explain the mechanism, which arises the local deformation inhomogenities in the areas of flaw, FEM calculations and experiments were performed on defects in adhesive bonds. The investigated specimen was prepared by two sheets of aluminum overlapped bonded with an artificial defect in the adhesive layer, fig. 18. The specimen was heated by a foil, which was fixed on the rear metal sheet. Although this is a severe simplification of the problem, it is possible to simulate the effect of thermal load on components with internal flaws. The calculated temperature distribution shows a inhomogenity above the internal disbonded area, fig. 19. The temperature difference between the maximum and the surrounding area is about 2/1000 K. This small difference cannot be detected by means of thermographic cameras. However, the change in thermal induced deformations is clearly visible, fig. 20. But there was nearly no change of the deformation whether the pressure of the air inside the defect was taken into account or not. This means, that in thermal loading the inhomogenity of the temperature distribution is the main effect for the deformation and this load enables to detect flaws even if there is nearly no air included.

6. Summary

The experiments with CFRP components proved the holographic interferometry to be a reliable in optimization of components and in detecting material defects. As an example, it was possible to detect stringer separations down to a length of 20 mm and delaminations down to a lateral length of 10 mm in a more than 10 mm thick CFRP wall of an aircraft component.

The measurements on a real large Airbus aileron proved the ability to use this method in industrial environment. An adapted development should arise an easy to use equipment for fast and reliable test in production line and in perodic quality tests.

The deformation measurements on antenna panels demonstrated the advantage of holographic interferometry for construction optimization. Some of the measurements, e.g. the gravity experiments, could only be performed with a non-contacting optical method.

7. Acknowledgements

We want to thank our coworkers Thomas Kreis and Jürgen Geldmacher for their excellent work in the different parts of the research work.

We want to thank both the German Research Association DFG and the BMFT German Minister for Research and Technology for their sponsorship of the research and we hope that we can finish the work of direct calculation of internal flaws. Then holographic interferometry is the real tool for defect quantification in HNDT.

8. References

/1/ R.L. Powell, K.A. Stetson, "Interferometric Vibration Analysis by Wavefront Reconstructions", J.Opt.Soc.Am. 55,1593-1598 (1965)
/2/ R. Dändliker, B. Ineichen, F.M. Mottier, "High Resolution Interferometry by Electronic Phase Measurement", Opt. Commun. 9 (4), 412-416, 1973

/3/ W. Jüptner, "Automatisierte Auswertung holografischer Interferogramme mit dem Zeilen-Scanverfahren", Proc. Frühjahrsschule '78, Hannover 1978
/4/ Th. Kreis, W. Jüptner, "Digital Processing of Holographic Interference Patterns Using Fourier-Transform Methods", Proc. IMEKO-Symp. on Laser Applic. in Prec. Measurements, Budapest, Plenum Press Ltd., London 1987
/5/ W. Jüptner, "Holographic Interferometry as Means for Nondestructive Testing and Flaw Quantization", SPIE Proc. 699 Laser and Opto-Electronic Technology in Industry: Stat-of the Art Review, 88-93, Xiamen, China 1986
/6/ H. Kreitlow, Th. Kreis, W. Jüptner, "Holographic interferometry with reference beams modulated by the object motion", Appl. Opt., vol.29, no.19, 4256-4262, 1987

Figures

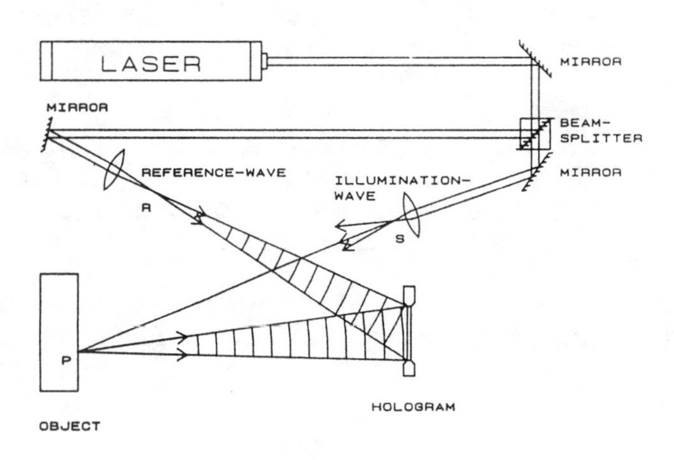

Fig. 1: Holographic storage of the optical wavefront

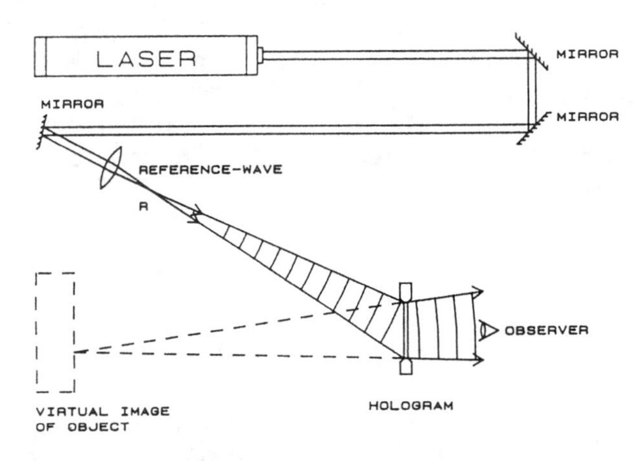

Fig. 2.: Holographic reconstruction of the optical wavefront

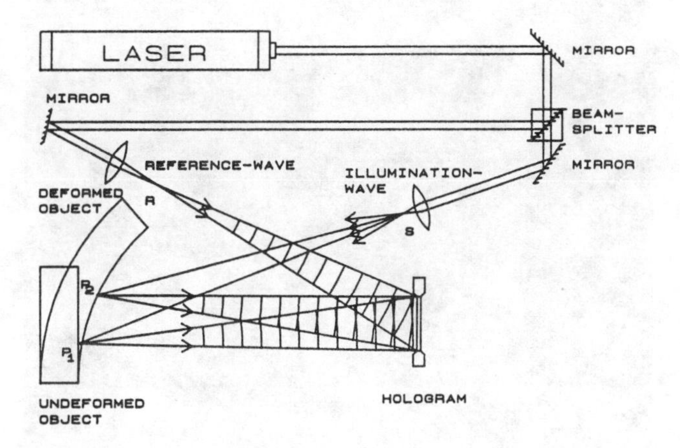

Fig. 3: Holographic recording of double exposure interferograms

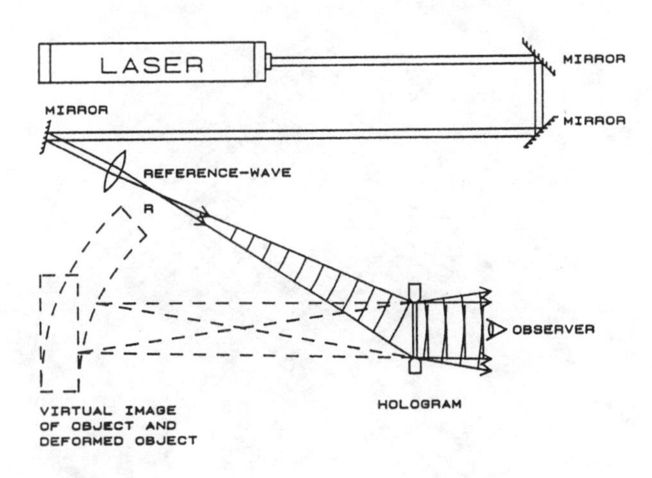

Fig. 4: Reconstruction of holographic fringe patterns

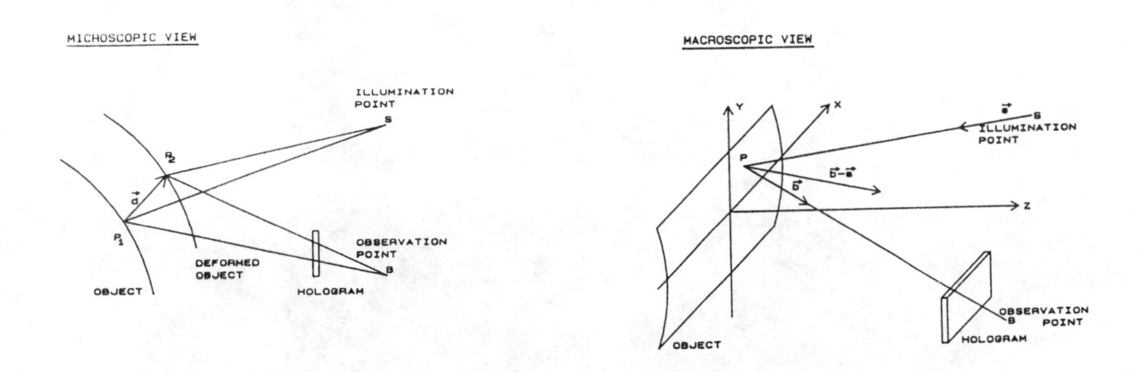

Fig. 5: Intensity distribution in holographic interferometry

Fig. 6: Holographic interferometer for antenna panel optimization

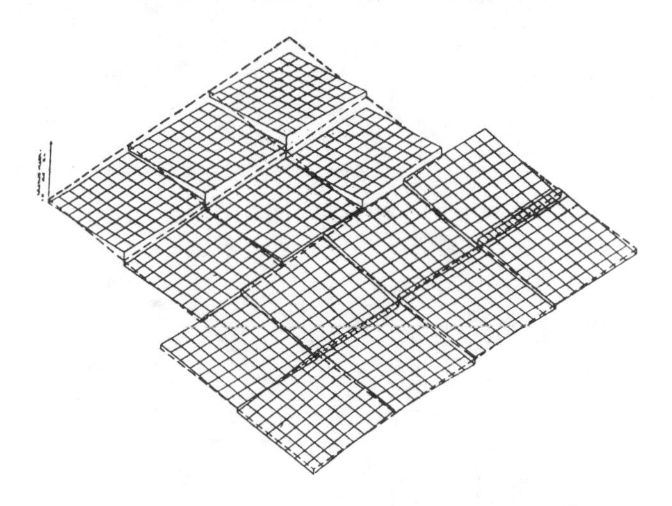

Fig. 7: Deformation measurement for a complex structure

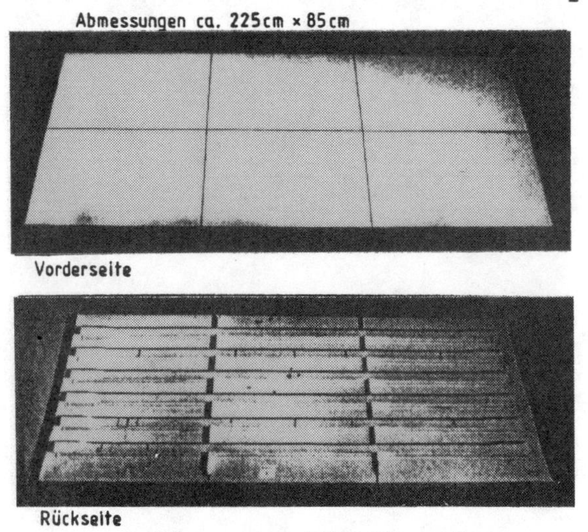

Fig. 8: CFK-Panel

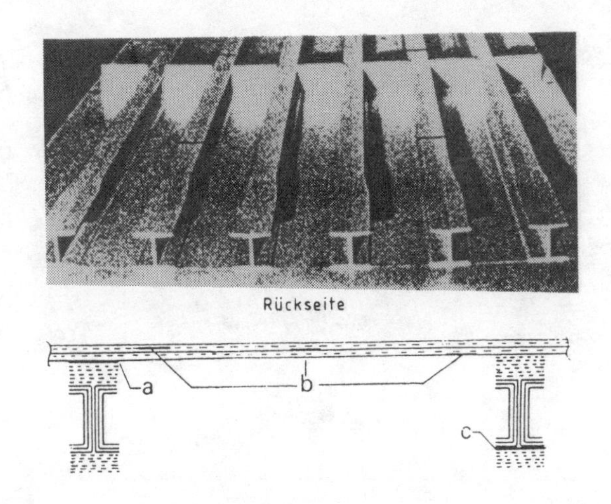

Fig. 9: CFK-Panel with artificial defect

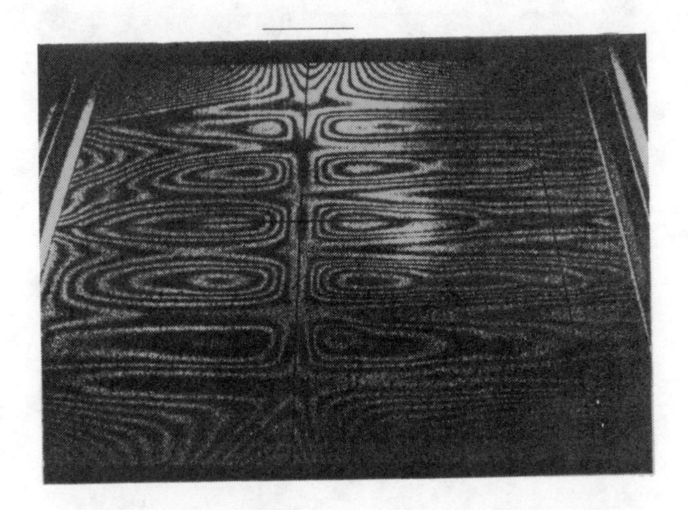

Fig. 10: Holographic fringe pattern under thermal load without defect

Fig. 11: Interferogram under thermal load with stringer-separation

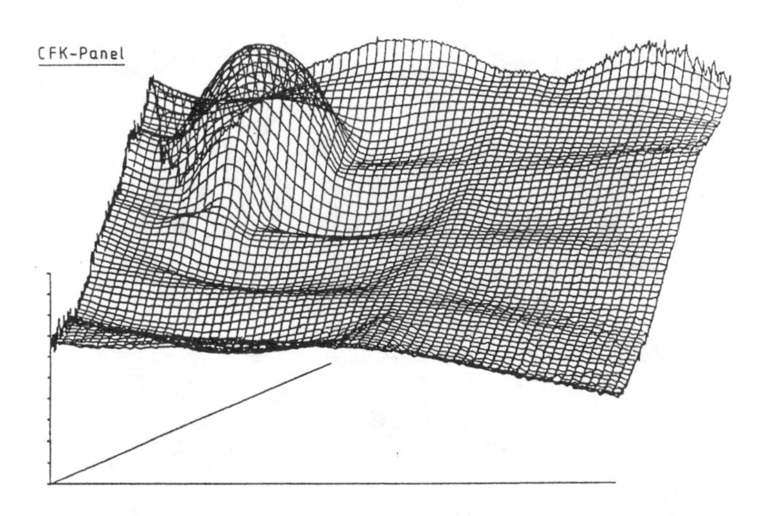

Fig. 12: Evaluation of the deformation field based on the interferogram

Fig. 13: Interferogram under thermal load with delamination

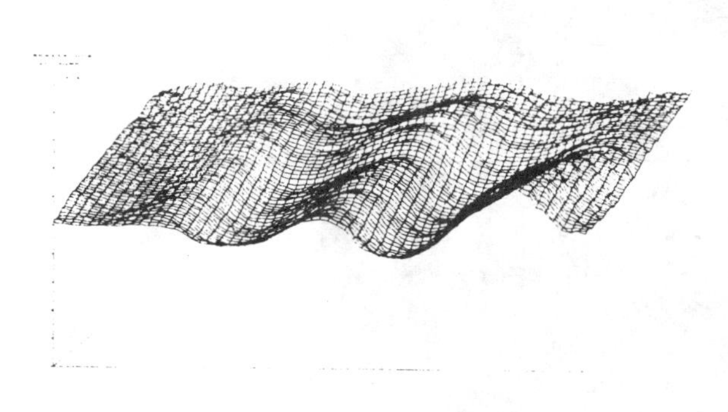

quasi-3-d

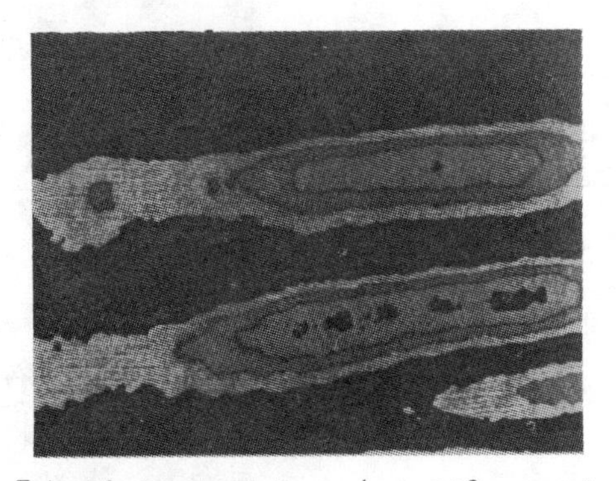

Falschfarbendarstellung (um 90° gedreht)

Fig. 14: Evaluated deformation field of an Airbus rudder

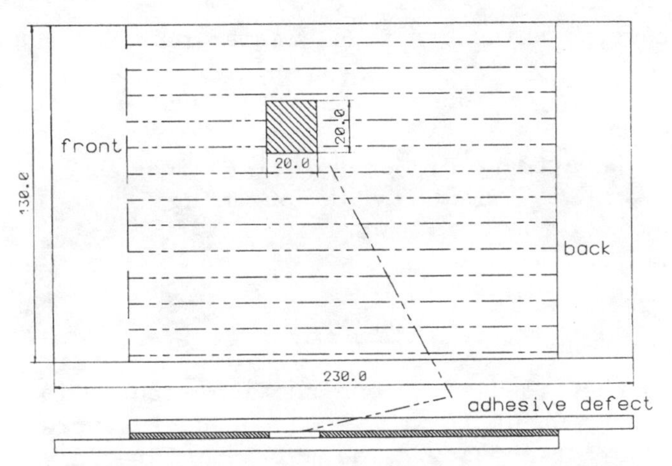

Fig. 15: Overlap adhesive bond

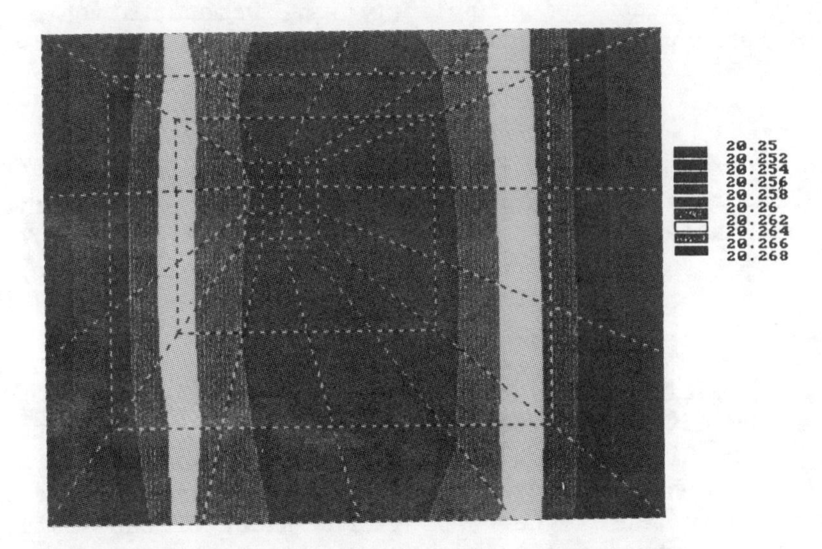

Fig. 16: Temperature distribution on the front side of the specimen

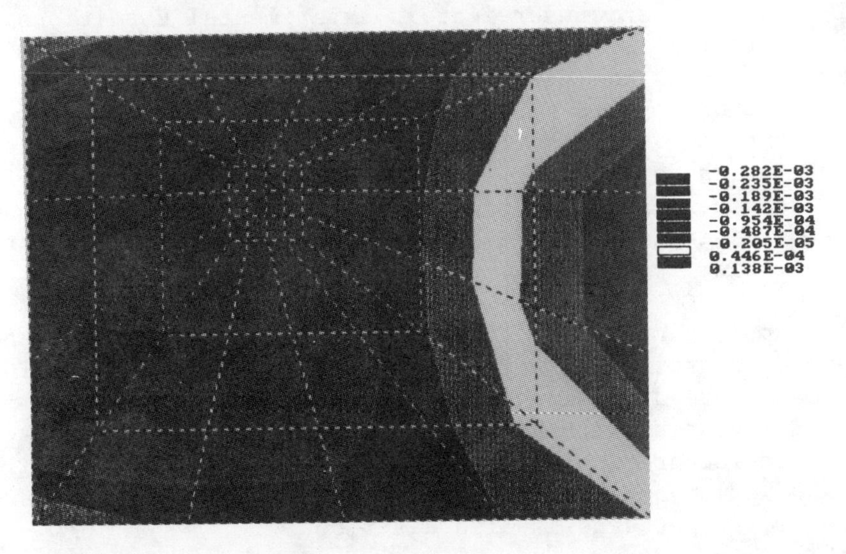

Fig.17: Deformation distribution on the front side of the specimen

Contouring using gratings created on a LCD panel

A. Asundi and C.M. Wong

Department of Mechanical Engineering
University of Hong Kong
Hong Kong

ABSTRACT

The shadow moire method is used in conjunction with a Liquid Crystal Display (LCD) projection panel onto which is written a computer generated grating to display contours of the topography of objects. The ability of rapidly changing the parameters of the grating like the pitch and width of lines through the software makes this a versatile tool for shadow moire applications. Demonstration of the technique including means for rapid image enhancement are described in this paper.

1. INTRODUCTION

The shadow moire method (1) consists of observing the moire resulting from the superposition of a reference grating and the shadow of this grating wrapped around the object. This shadow grating which conforms to the shape of the object acts as the deformed grating and the moire pattern thus maps the topography of the object. Various applications of this technique have been developed (2-5) with the majority of them being in the medical field.

A typical schematic of the shadow moire set-up is shown in Fig. 1. A parallel beam of light illuminates the grating which is placed adjacent to the object, at an angle 'α' and the resulting moire pattern is observed in the normal direction. In this instance topographic contours are seen in the form of moire fringes where the contour interval is ($p/\tan \alpha$) where 'p' is the pitch of the grating. In general to contour large objects where collimation of the light beam is impractical, diverging light can be used. Similar contours result although the contour interval need not be constant (1, 2). However, under certain experimental conditions, identical relations accrue. The contour interval can be made smaller, i.e. the sensitivity can be increased by decreasing the pitch of the grating 'p', or by increasing the angle of incidence 'α'. Changing the pitch of the grating requires numerous gratings of different pitch to be available. Asundi (6) has proposed an alternative which involves rotating the grating in its plane, thus altering the effective pitch continuously. On the other hand the angle of incidence can be continuously varied to change the sensitivity. Both these approaches require moving parts which might not always be desirable.

In this paper, a Liquid Crystal Display (LCD) projection panel onto which a computer generated grating is written is used as the grating. The computer generated grating can be readily changed in respect of its pitch and width of white line, thus providing a grating with variable pitch and no moving parts. This paper describes the implementation of this LCD panel and demonstrates its versatility for contouring. Other features of the panel, viz. the contrast reversal switch used for enhancing the moire patterns is also described.

2. EXPERIMENTAL DETAILS

With the widespread use of computers the need to disseminate computer generated information to a large audience has led to the development of the Liquid Crystal Display projection panel. This panel consists of a two-dimensional array of elements (640 x 400) in a display area of 231.1 mm x 144.8 mm. Each of the picture elements (pixels) can be individually addressed to switch it on (bright) or off (dark). Indeed the elements could be partially turned on in eight steps thus generating 8 gray levels. The panel is connected to the computer via the video port and is compatible with the CGA/EGA graphics cards. The panel can thus display information as if it were the PC monitor and being transparent can be placed on an overhead projector for projecting to a large audience.

For this particular application the set-up is slightly different in that the LCD panel is not used for projection display but rather as a liquid crystal venetian blind which filters the light passing through. This feature is readily obtained by generating a computer grating. A grating is a series of dark and bright lines of generally equal width and the spacing between adjacent dark (or bright) lines called the pitch being generally constant over the whole field. This grating with user selectable pitch can be software generated and displayed on the monitor as shown in Fig. 2. The pixels corresponding to the dark lines are off while those corresponding to the bright lines are on. This grating when displayed on a transparent LCD panel forms the grating for the shadow moire experiment. The schematic of the set-up is shown in Fig. 3 and is similar to that of Fig. 1. Shadow moire fringes are readily seen on the object when viewed through this LCD panel as in Fig. 3. The sensitivity can be easily changed just by selecting a grating of different pitch as evidenced in the lower photograph of Fig. 3. The fringe contrast is quite good and optimum fringe visibility can be obtained by adjusting the contrast on the LCD panel.

The versatility of the LCD based shadow moire method is demonstrated using a hemispherical object welded around the equator and painted white for good contrast shadows. An alternative to conventional imaging, is digital imaging which displays the picture to be recorded in real time and is thus adopted in this method. In fact a software program is developed which controls both the LCD panel and the digital imaging facility. The whole system operates on PC-AT with the frame grabber slotted in. The routines described herein are in general monadic, point by point operations and are thus performed very quickly. The image as acquired by a CCD camera and digitized by the frame grabber before display has a relatively poorer contrast than those recorded using a 35 mm camera. One rapid means of enhancing the contrast is to use a moving average thresholding (MAT) where the pixel of interest (POI) is bright or dark depending on its value as compared to the average of pixels ahead of and behind the POI. This thresholding scheme is quite fast and compensates for non-uniformity of illumination. However for best results the grating lines should be normal to the scanning direction which in this case is incidental. Fig. 4 shows the thresholded shadow moire patterns for different pitch of grating while Fig. 5 are the patterns for the same pitch but different ratios of black to white line widths. These patterns were recorded directly of the display monitor. To get rid of the grating lines addition of the two images with contrast reversal of the LCD between the exposures would not work since the images are thresholded and the intensity variation between the dark and bright fringes will be barely noticeable. An alternate to adding the two images is to AND the two images. Since the AND of two one bit binary numbers is 1 only if the two numbers are initially 1 also, else it is 0, thus the bright moire fringes would remain bright while all other points would be dark. Fig. 6 (a) and (c) show that this is indeed the case. These two patterns were recorded using two different CCD cameras and thus the difference in contrast of the fringes in the two cases. The logical exclusive OR (XOR) is another way of accomplishing the same goal. The XOR operations between two one-bit binary numbers is similar to the AND operation with the exception that the XOR of (0, 0) is 1 while the AND would give a zero result. Fig. 6 (b) and (d) testify to this with the grating lines almost completely obliterated. The fringe patterns nevertheless are still not upto par.

Thus another approach to fringe enhancement is developed. Once again full frame processing methods are utilised to speed up the enhancement routines. The procedure adopted in this instance is as follows:

(a) Obtain the shadow moire pattern corresponding to a particular pitch 'p' of the LCD grating. This is picture 'A' as shown in Fig. 7 (a).
(b) Reverse the contrast of the LCD grating to give picture 'B'.
(c) Obtain picture C = (A-B) and stretch the contrast of C using linear contrast enhancement with saturation. This is conveniently done using Look-Up Tables (LUT). The result is shown in Fig. 7 (b). Note that this contrast enhancement affects only the display and not the buffer stored values.
(d) Obtain picture D = (B-A).
(e) Take the XOR of C and D and the pattern in Fig. 7 (c) shows the enhanced shadow moire fringes.

This method is best suited for an LCD panel where reversal of contrast is accurately and easily achieved. This would be very difficult with actual gratings which would require precise shift of half a pitch of the grating.

3. CONCLUSION

Contouring using grating generated on a Liquid Crystal Display (LCD) projection panel adapted for shadow moire experimentation is demonstrated. Advantages of this method which include ability to rapidly change the pitch of the grating and the ratio of the widths of the black to white lines are exemplified. Elimination of the generating grating lines can be simply achieved by a double exposure technique. However for digital images double exposure is achieved through logical operations. Digital fringe enhancement routines are also described using quasi-real-time methods for excellent fringe contrast enhancement.

4. ACKNOWLEDGEMENT

The support of the K.K. Leung Research and Teaching Endowment Fund and the Committee for Research and Conference Grants is gratefully acknowledged.

5. REFERENCES

1. F.P. Chiang, "Moire Methods of Strain Analysis", Manual on Experimental Stress Analysis, ed. A.S. Kobayashi, Chap. 6, SESA 1978.
2. H. Takasaki, "Moire Topography", Appl. Opt., Vol. 12(4), pp. 845-850, 1973.
3. H. Takaski, "Moire Topography", Moire Applied to Medical Applications, ed G. von Bally, pp. 45-59, Springer Verlag, 1979.

4. A. Asundi and C.M. Wong, "Pedemoirography", Proc. SEM Spring Conference New Orleans, 1986.
5. A. Asundi, "An Optical Foot Pressure Measurement Technique", Biomedical Engineering V: Recent developments, pp. 321-324, Pergamon Press, 1986.
6. A. Asundi, "Variable Sensitivity Shadow Moire Method", Jnl. of Strain Analysis, Vol. 20(1), pp. 59-61, 1985.

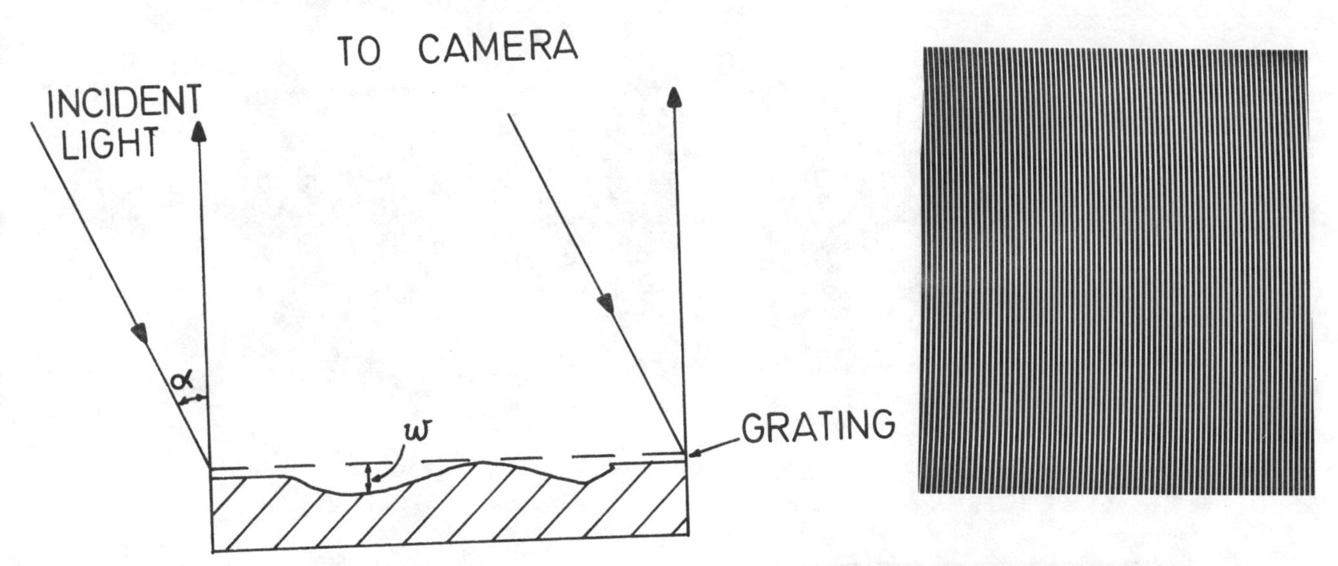

Fig. 1. Shadow moire principle.

Fig. 2. Computer grating.

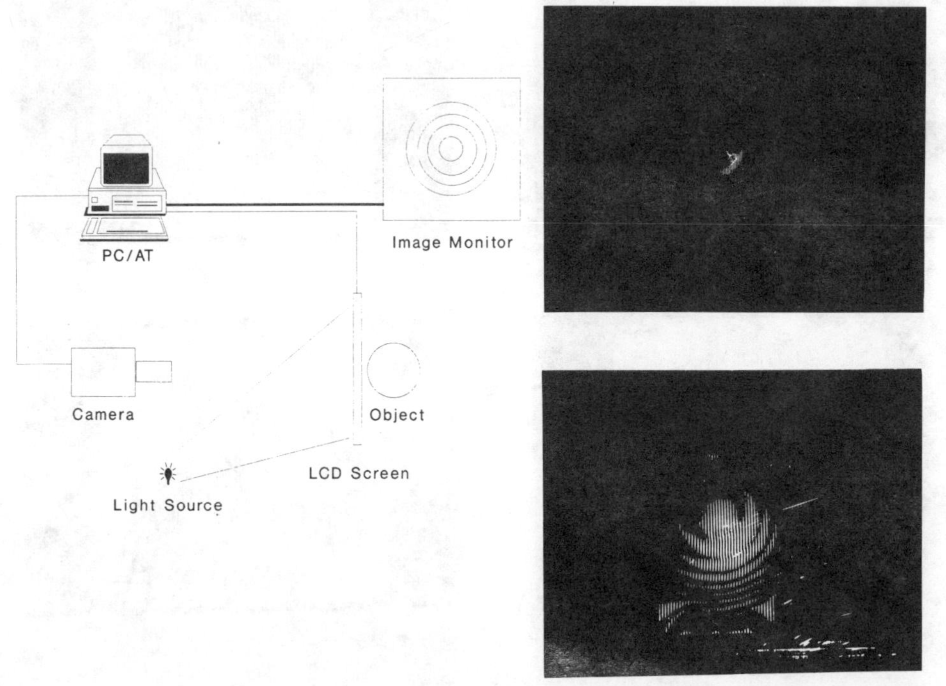

Fig. 3. Schematic and the moire patterns using the LCD based shadow moire method.

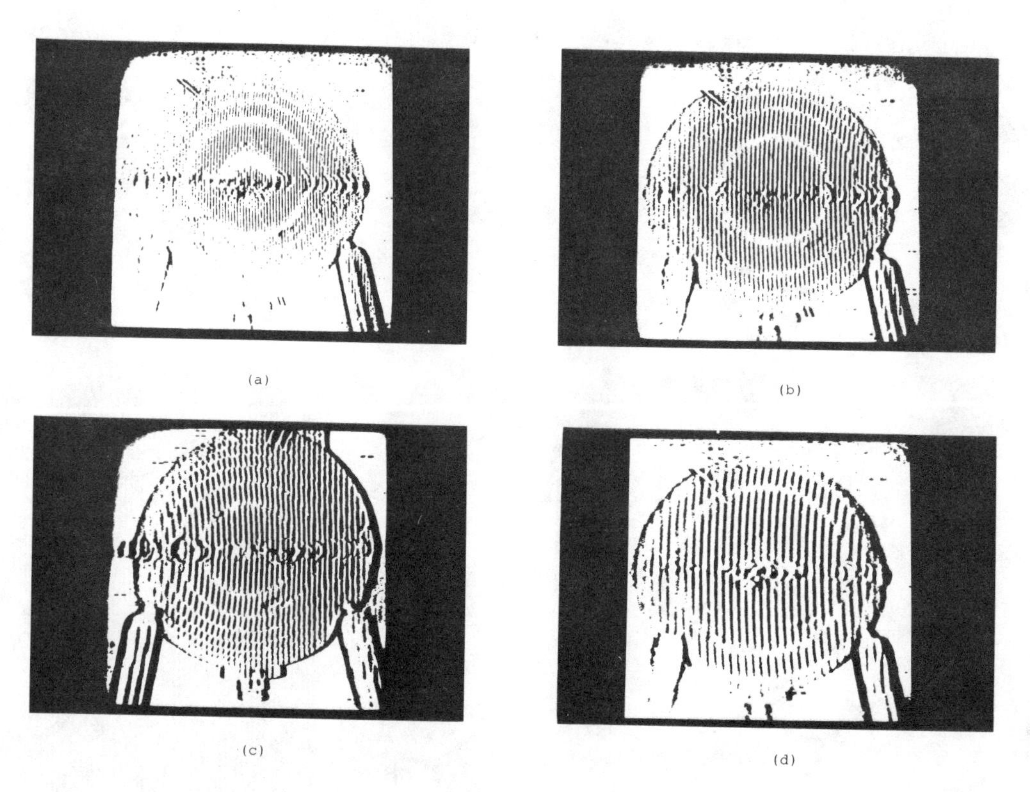

Fig. 4. Shadow moire contours of spherical ball welded along the equator for different pitch of LCD grating.

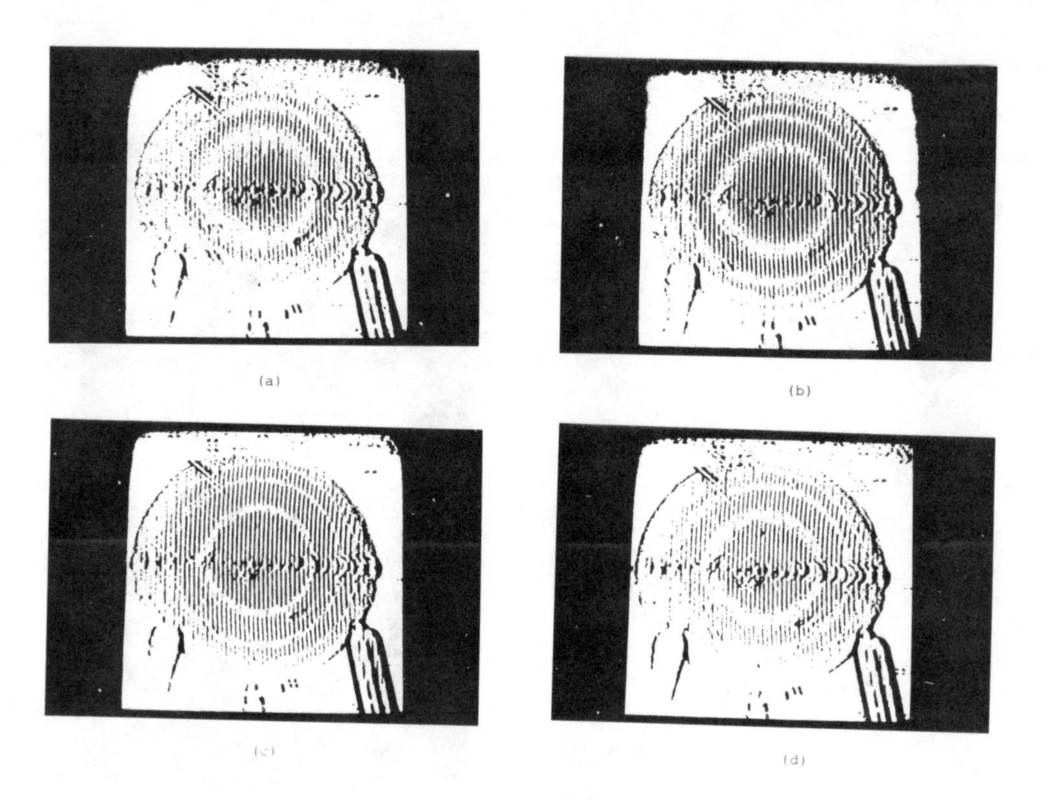

Fig. 5. Shadow moire contours with same pitch but different bright to dark line width ratio.

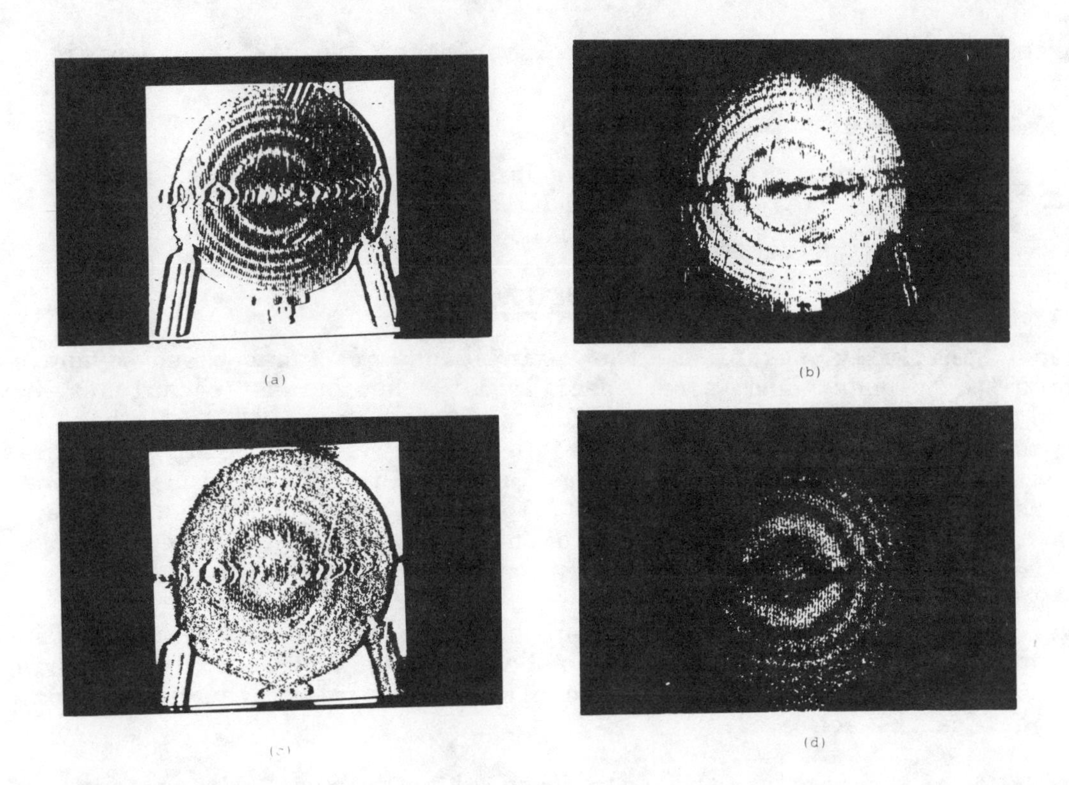

Fig. 6. Fringe enhancement using logical operations on Shadow moire patterns with complementary gratings. (a) and (c) AND operation (b) and (d) XOR operation.

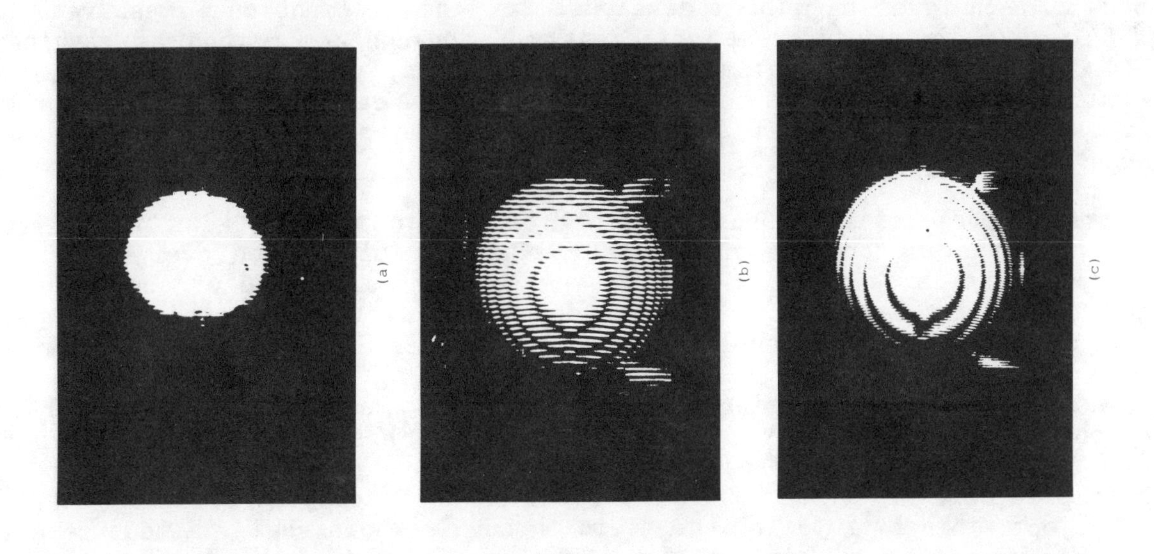

Fig. 7. Fringe enhancement scheme using whole frame operation. See text for details.

Three-dimensional inspection using laser-based dynamic fringe projection

D.M. Harvey, M.M. Shaw, C.A.Hobson, C.M. Wood, J.T. Atkinson, M.J. Lalor

Coherent and Electro-Optics Research Group,
Liverpool Polytechnic, Byrom Street,
Liverpool, L3 3AF, U.K.

ABSTRACT

This paper initially describes the principles of laser-based dynamic fringe projection techniques, and more specifically, how these techniques have been exploited in the Dynamic Automated Range Transducer (DART). This instrument employs novel techniques to produce a high accuracy single point range measurement over a wide variety of distances. The latest miniature optical components lend themselves to the production of a small, low-cost, self-contained device, which is easily portable and suitable for use in on-line, automated inspection systems, or in the field.

The development of the DART for three-dimensional measurements is described. Two methods are possible, either a single beam or twin-beam system. Using a CCD photodiode array with precise pixel geometry an accurate three dimensional mapping of an object can be achieved.

The problems of processing the data from the various DART systems is briefly examined. For multi-pixel measurements a parallel architecture using fast digital signal processing devices is recommended.

1. INTRODUCTION TO LASER-BASED THREE DIMENSIONAL GAUGING

A number of techniques have been developed for the 3-dimensional measurement of shape. To take the measurement efficiently a non-contact method of gauging is required. This allows the automation of the process, yielding many applications in industrial situations. The coupling of the optical non-contact measurement machine, with suitable optical sensors interfaced to a modern computer, enables on-line process monitoring to be achieved.

Static fringe projection methods have been used[1] to measure wear in surfaces. However these systems produce only relative height information over the surface and are ambiguous in determining whether the surface is concave or convex. Phase stepping techniques based on holographic or Moiré systems can solve the height ambiguity problem[2] but their application in an industrial environment is limited.

A new novel approach to single point distance measurements using dynamic projection was produced at Liverpool Polytechnic[3]. This technique has since been expanded to produce 3-dimensional information about objects under inspection[4,5]. The basic principle of dynamic fringe projection is to project a field of fringes onto an object, then by rotating the field through a known angle, a measurement of the absolute range of any point on the object can be taken.

2. DYNAMIC AUTOMATED RANGE TRANSDUCER (DART)

Figure 1 shows a general arrangement for gauging the distance Z of an object from a cosinusoidal grating. The grating is placed between a point source and the object, projecting a shadow of the grating into the object area. The particular ray, which strikes the object, passes through the grating at point Y_g. The distance Z is related to the number of bright fringes N, which cross the optical axis beween the object and the grating.

The range Z can be shown to be[6]:-

$$Z = \frac{Z_f N P}{Y_s - N P} \qquad (1)$$

where P = period of the grating at a particular position in space.

Counting fringes in space is an unsatisfactory arrangement for determining the fringe order N. Automation of the fringe counting is achieved by rotating the grating about the optical axis. This varies the period P of the grating according to the angle of rotation Θ

$$\text{where:- } P = \frac{Po}{\sin \Theta}$$

and Po the period of the grating is approximately 1mm.

The range can be computed automatically from:-

$$Z = \frac{Z_f N P_o}{Y_s \sin\Theta - N P_o} \qquad (2)$$

where N is real number corresponding to the fringe order at an angle Θ.

A DART system for single point distance measurements figure 2, has been evaluated. A computer controlled stepper motor rotates the grating via a gearing system attached to the outside of the grating. An optical shaft encoder provides accurate positional information of 10,000 pulses per revolution. With suitable gearing the position of the grating can then be determined in the milli-radian range.

Figure 3 shows the optical arrangement of the DART for a single point range measurement.

The intensity of the detected optical signal from a point on the surface has been shown to be[6]:-

$$I(Z) = \cos^4 (N\pi\sin\Theta) \qquad (3)$$

Figure 4 shows a typical received signal where the fringe order N = 5.

The DART gives the fringe order by counting the number of peaks and fractions of a peak as the grating rotates from 0 to 90°. As the signal is symmetrical about 90° (1.57 radians on figure 4) the rotation of the grating from 90° to 360° can be used for computer processing to calculate a value for the distance Z.

Techniques have been developed for processing the received intensity and positional signals. Accuracies of > 99% have been achieved over distance measurements of 10-250mm.

Miniaturisation of the DART is presently being investigated. The use of a laser diode with miniature optical components should enable a hand-held detection head to be produced. By linking this device to an on-board computer, spot length measurements should be possible.

3. MULTI-POINT DART

The novel techniques employed in the dart can be extended to perform multi-point measurements. An array of samples taken over the surface of an object using the DART, allow a three dimensional measurement of the object shape.

3.1 Off-axis range measurement

The principles of angular based range measurement developed in the DART uses a single laser beam to accurately measure the distance of a single point.

However the measurements so far considered are of points located on the optical axis. For points off the optical axis their range can be determined as in figure 5. The range of the two points can be calculated from:-

$$Z1 = \frac{Z_f N1\ P_o}{(Y_s + X)\sin\theta - N1\ P_o} \tag{4}$$

$$Z2 = \frac{Z_f\ N2\ P_o}{(Y_s - L)\sin\theta - N2\ P_o} \tag{5}$$

where:- N1 and N2 are the fringe orders at Point 1 and Point 2 respectively,
X and L are the distances of the respective pixels from the optical axis.

The determination of range of an off axis point using this technique requires both the fringe order, and the distance of the respective pixel from the optical axis. This makes a 3-dimensional mapping of the object possible by evaluating a distance Z for each x-y pixel of the object, imaged onto the detector.

3.2 Two-point dynamic fringe projection

A second technique has been developed to measure range which only requires the fringe order to be determined to produce a 3-dimensional multi-pixel map of the project. By using two laser beams projecting fringes onto the object from two different angles it is possible to make off axis range measurements (figure 6). The fringe order of the intensity signals from the two point sources must however be determined.

The light from the two lasers is spatially filtered to effectively form two point sources at distances Z_f from the rotating grating. These light sources can be chopped by electronic shutters to illuminate the object when required. When using laser light sources only one shutter would normally be open at any particular time, to reduce any possibility of the two beams producing interference patterns on the object. The effects of any interference patterns however will be minimal because the fringe spacing Po of the grating is large compared to the wavelength of the laser light. Workers at The Liverpool Polytechnic have investigated 3-D measurements using two beam dynamic fringe projection. For this work if interferences phenomena create measurement difficulties, the laser light sources are replaced by strong white light sources. This removes any problems associated with coherent light interference, but still allows the shadow of the grating to be clearly projected onto the object.

The range of point P can be evaluated from[5]:-

$$Z = \frac{Z_f P_o (N1 + N2)}{2Y_s \sin\Theta - P_o (N1 + N2)} \qquad (6)$$

where:- N1 and N2 are the fringe orders from point sources associated with Laser 1 and Laser 2 respectively.

The accuracy of the multi-point DART should be comparable with the single-point DART, allowing 3-dimensional volume and surface measurements to be made effectively.

3.3 Suitable detectors

Initial measurements on a prototype system have been made using photomultipliers to detect a single point measurement[3]. A prototype engineered DART system[6] using photodiode detectors has been successfully commissioned, using the detector as the object to increase the optical signal received from the system. Positioning the detector on the optical axis to receive light dispersed from the surface of interest (as figure 6) reduces the signal by a factor dependent on the properties of the surface. To counteract this reduction in signal either a more powerful light source or a more sensitive detector can be used.

To obtain three dimensional information about a number of points on the surface of the object under inspection a two-dimensional detector is required. Suitable CCD photodiode array detectors are becoming available. The sensitivity of CCD photodiode detectors to visible light is improving, making the latest devices appropriate for use in the multi-pixel recording of dynamically projected fringes.

4. SINGLE POINT DATA PROCESSING

The data collected for a single point measurement using the DART requires processing into a range. Various methods have been tried to determine the fringe order N. Measuring the number of zero crossings is fast, but determination of any fractional fringes is difficult. The best accuracy is achieved by evaluating the fringe order as the grating rotates from 0 to 90°. This is due to the fringe around 90° having an extended period due to the turnover of the grating. Curve fitting techniques best determine the real fringe order N but processing times are lengthy. Initial apparatus used Intel 8086/87 and Transputer based signal processors, which were sufficiently powerful when computing single point measurements.

4.1 Multi-pixel data processing

The data collected from a multi-point measurement is enormous. A CCD area detector collecting fringe information at each pixel on its array would produce large amounts of data from the surface under inspection. To process each pixel fast parallel computer architectures are being investigated. Depending upon the processing performed by each processor either a low level systolic array of semi-custom integrated circuits or a higher level parallel system using proprietary digital signal processors is proposed. The Texas TMS320 series of digital signal processors[7] are currently being investigated. They allow flexibility to be built into a system since they are reprogrammable. The latest devices offer fast processing speeds and good communication facilities between devices, making them ideal for parallel processing applications.

5. CONCLUSION

The DART is a novel device for the absolute measurement of range. Its development as either a single point or multi-point mesurement apparatus has been described. The use of dynamic fringe projection is fundamental to the operation of the DART. Suitable detectors for the recording of fringes dynamically projected onto an object have been discussed. The evaluation of range by an on-line computer has been briefly considered for both single pixel and multi-pixel measurements.

The accuracy of a single point measurement of > 99% over distances up to 250 mm has been achieved. Similar accuracies for three-dimensional measurements should be possible.

Fast data processing techniques have been briefly discussed. The use of a parallel architecture for multi-pixel measurements is recommended using the latest version of a family of digital signal processing devices.

6. REFERENCES

1. M.B. Koukash, C.A. Hobson, M.J. Lalor, J.T. Atkinson, "Detection and Measurement of Automatic Fringe Analysis", Automatic Fringe Analysis, Ed. B.L. Button, G.T. Reid, pp.97-107, FASIG, Loughborough, 1986.

2. G.T. Reid, R.C. Rixon, H.I. Messer, "Asolute and Comparative Measurements of 3D Shape by Phase Measuring Moiré Topography", Optics and Laser Technology, Vol.16, No.6pp.315-319, 1984.

3. J.T. Atkinson, M.J. Lalor, "A novel approach to optical range finding", Proc. SPIE, Vol.701, pp 237-243, 1986.

4. M.M. Shaw, D.M. Harvey, C.A. Hobson, M.J. Lalor, "Non-contact ranging using dynamic fringe projection", Proc. SPIE, Vol.1163, pp 22-29, 1989.

5. C.M. Wood, M.M. Shaw, D.M. Harvey, C.A. Hobson, M.J. Lalor, J.T. Atkinson, "Absolute range measurement system for real-time, 3-D vision", Proc. SPIE, Vol.1332, 1990.

6. M.M. Shaw, J.T. Atkinson, D.M. Harvey, C.A. Hobson, M.J. Lalor, "Range measurement using dynamic fringe projection", Journal Physics D: Applied Physics, Vol.21, pp S4-S7, May 1988.

7. Texas Instruments, TMS 320C40 Data Sheet, 1990.

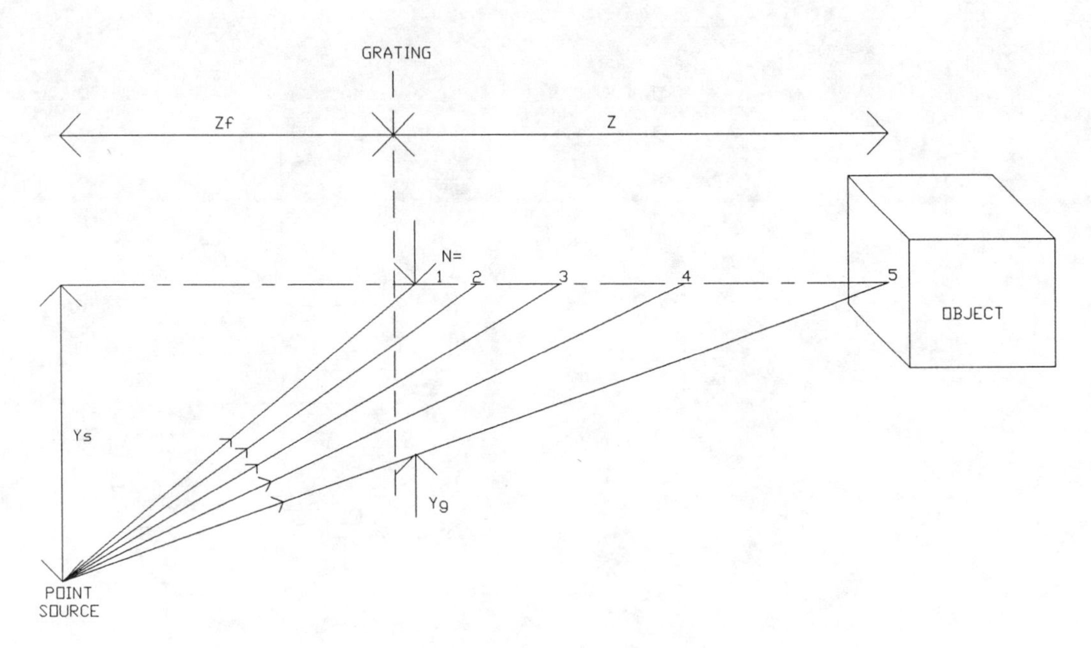

FIG.1. TRIANGULATION METHOD OF DISTANCE MEASUREMENT USING COSINUSOIDAL GRATING

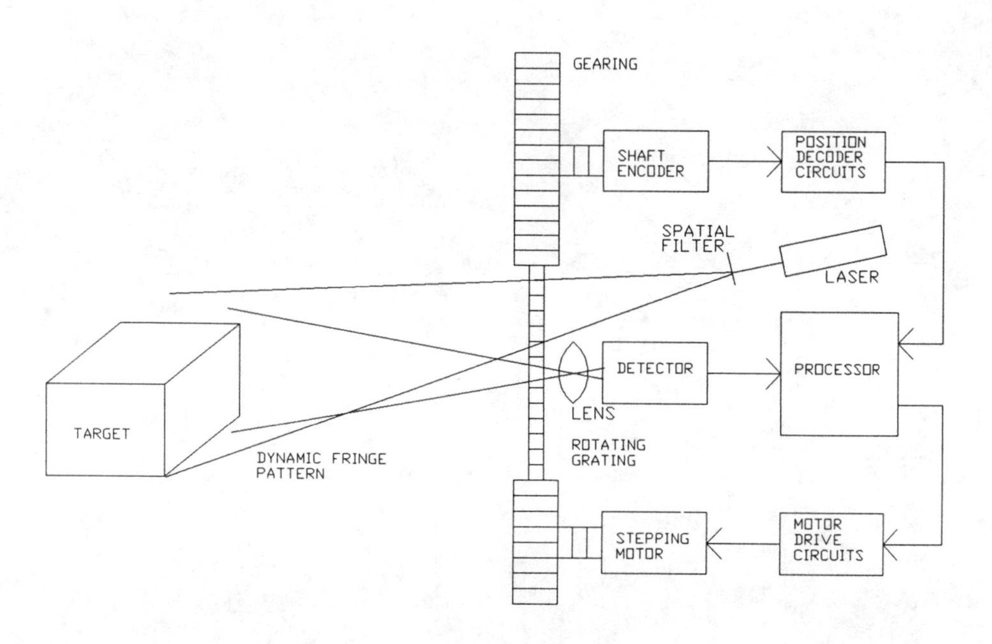

FIG.2. SCHEMATIC DIAGRAM OF DYNAMIC AUTOMATED RANGE TRANSDUCER (DART)

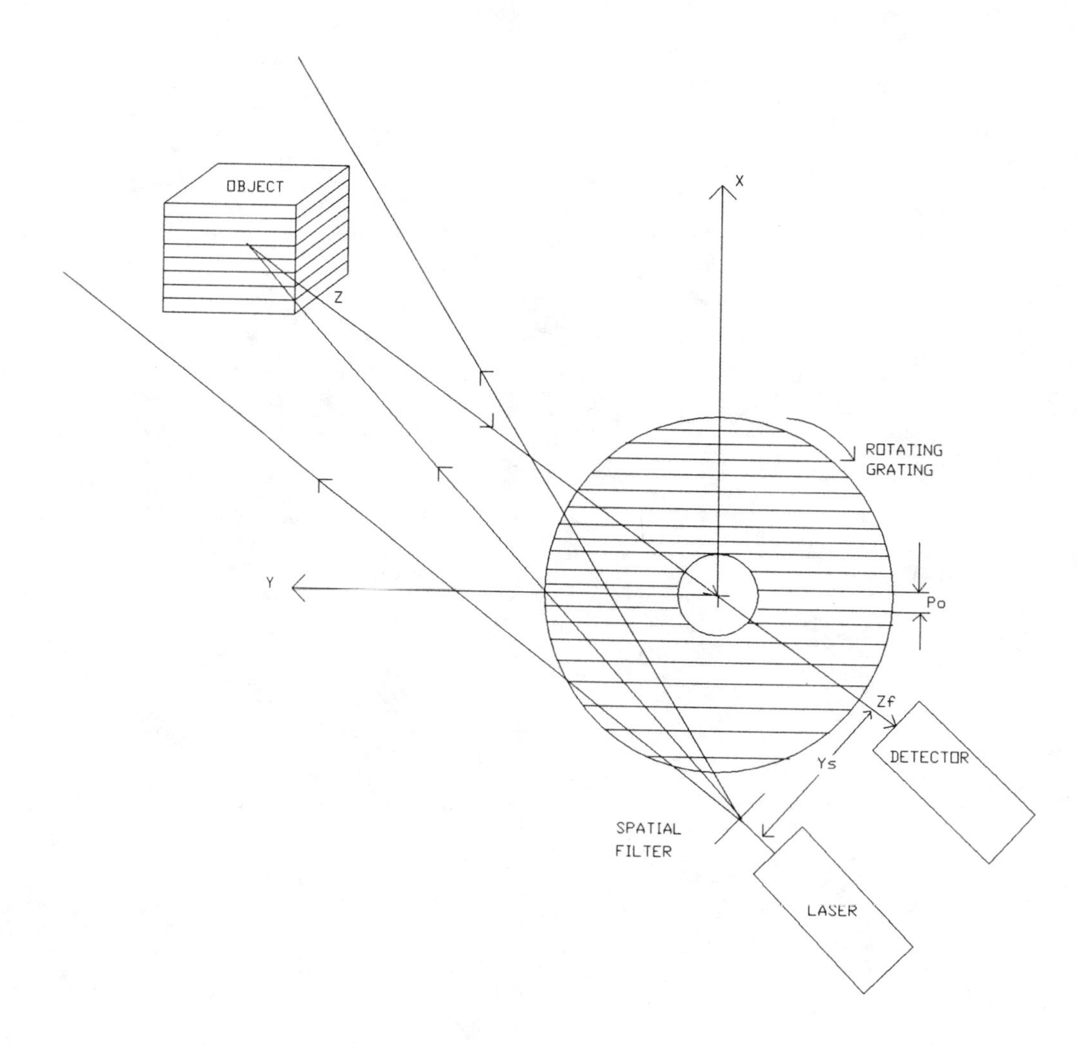

FIG.3. OPTICAL AND GEOMETRICAL ARRANGEMENT FOR SINGLE POINT DISTANCE MEASUREMENT

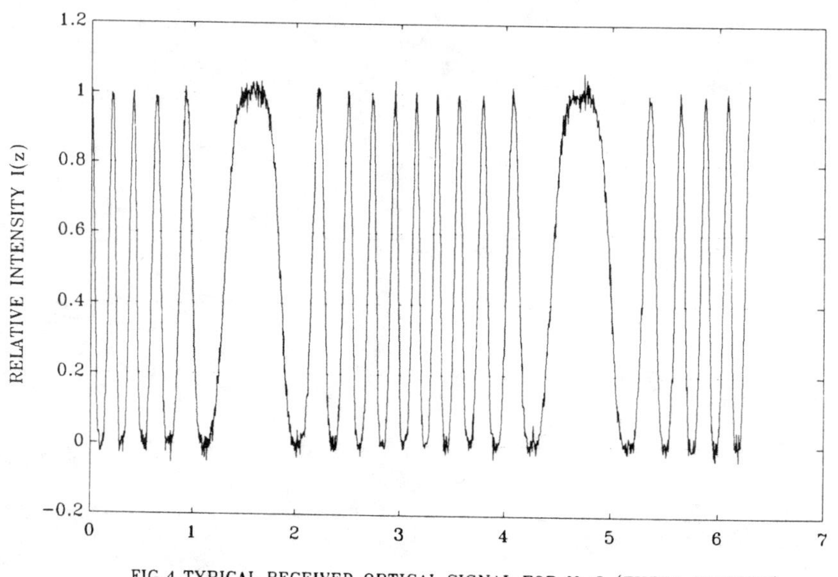

FIG.4.TYPICAL RECEIVED OPTICAL SIGNAL FOR N=5 (THETA RADIANS)

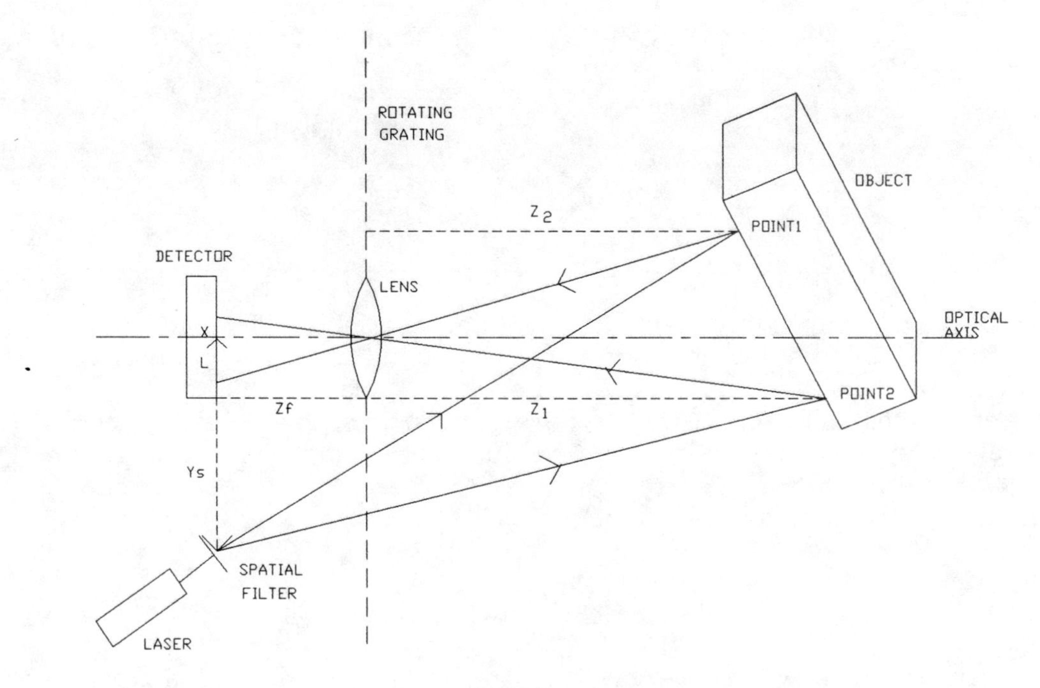

FIG.5. GEOMETRY FOR POINTS NOT ON OPTICAL AXIS

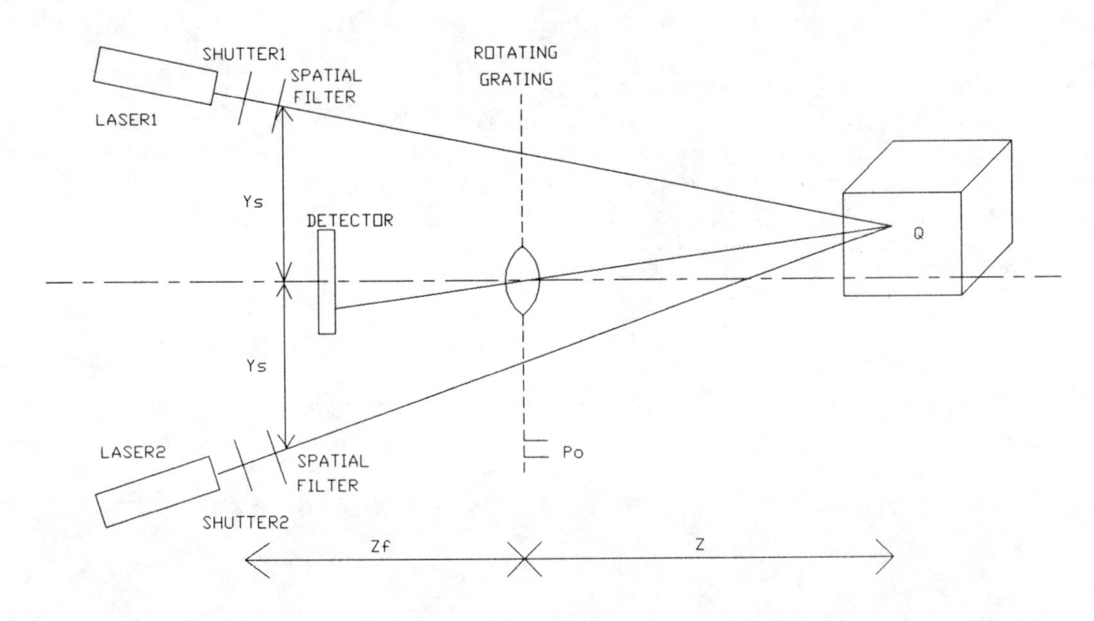

FIG.6. MULTI-PIXEL RANGE MAPPING DART

Method for evaluating displacement of objects using the Wigner distribution function

Joewono Widjaja, Jun Uozumi, and Toshimitsu Asakura

Research Institute of Applied Electricity, Hokkaido University
Sapporo, Hokkaido 060, Japan

ABSTRACT

A new method is proposed for evaluating the local displacement of objects in speckle photography by using the Wigner distribution function (WDF). It is applied to evaluate the displacement of one-dimensional (1-D) specklegrams as well as specklegrams of the ground glass and rubber plate which are slightly shifted and stretched, respectively. As a preliminary study, the local spectra $W(x,\omega)$ are qualitatively and quantitatively verified in comparison with the theoretical results and the conventional Young's interference fringes. The results show the feasibility of evaluating the displacement of objects using the WDF.

1. INTRODUCTION

In the study of speckle photography, an optical Fourier transform of the double exposed speckle pattern has been widely used to analyze displacement, which is caused by deformation or movement of objects, and by the flow of liquid. In most cases, such displacement is space-variant and requires a local analysis. Then we have to take many Fourier transforms of the specklegram regionally and repeatedly, since an average of the whole displacement is only the information that we can get from the Fourier transform operation. However, if we employ a space-frequency signal representation, we can obtain the local spatial frequency information in speckle photography, since it can display the spatial frequency as a function of position. Among space-frequency signal representations, the WDF has been found to be useful in many fields of science and engineering[1,2]. In this study, measurements of the 1-D displacement in 1-D specklegrams are discussed as a first step, because the WDF of a 2-D signal pattern is a 4-D function and is hardly realized optically[3-5]. Instead of it, the WDF of a 1-D signal pattern offers an easier performance of optical space-frequency image representations.

2. THEORETICAL BACKGROUND

The concept of the WDF was first introduced into optics by Bastiaans[6]. Its properties and relations with other space-frequency representations have been intensively discussed by Claasen and Mecklenbrauker[7].

The cross-WDF of 1-D signal patterns, $f(x)$ and $g(x)$, is given by

$$W_{f,g}(x,\omega)=\int_{-\infty}^{\infty} f(x+x'/2)\,g^{*}(x-x'/2)\,\exp(-i\omega x')\,dx'. \qquad (1)$$

It can also be defined from the Fourier transforms, $F(\omega)$ and $G(\omega)$, of $f(x)$ and $g(x)$, respectively, by

$$W_{F,G}(\omega,x)=(1/2\pi)\int_{-\infty}^{\infty} F(\omega+\omega'/2)\,G^{*}(\omega-\omega'/2)\,\exp(-ix\omega')\,d\omega'. \qquad (2)$$

The auto-WDF of a 1-D signal pattern $f(x)$ can be defined in two equivalent forms. The first form takes

$$W(x,\omega)=\int_{-\infty}^{\infty} h(x,x')\exp(-i\omega x')\,dx' \tag{3}$$

with the relation of

$$h(x,x')= f(x+x'/2)f^{*}(x-x'/2), \tag{4}$$

while the second form is given by

$$W(x,\omega)=\int_{-\infty}^{\infty} g(x,x',\omega)f^{*}(x-x'/2)\,dx' \tag{5}$$

with the relation of

$$g(x,x')= f(x+x'/2)\exp(-i\omega x'). \tag{6}$$

The first form of Eq.(3) can be interpreted as a combination of multiplication and Fourier transform processes, while the second form of Eq.(5) combines the two processes of multiplication and convolution. We find that the first form is somewhat easier to be implemented optically.

3. EXPERIMENTAL SETUP

The operation described by Eq.(6) can be realized optically by using the optical system shown in Fig.1[8]. The shifted funtion $f(x+x'/2)$ can be obtained by rotating a transparency object $f(x)$ with an angle θ and is determined by the following consideration:

$$f(x) \dashrightarrow f(xa\cos\theta + x'a\sin\theta). \tag{7}$$

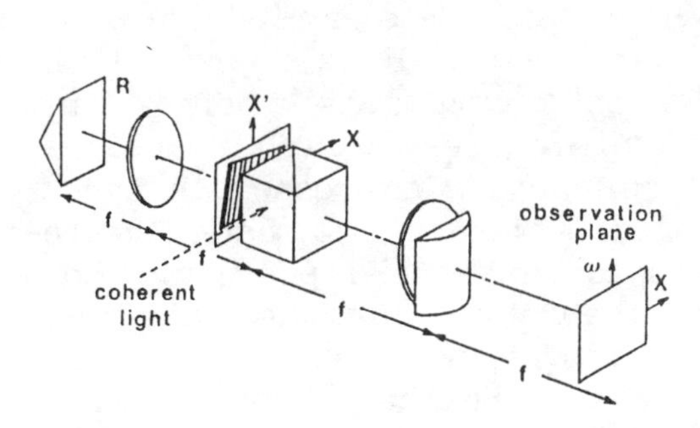

Fig.1. Optical system for producing the WDF of a 1-D object[8]. R = roof top prism.

When θ=26.6°, we obtain $a\cos\theta$=1 and $a\sin\theta$= 0.5 which generate the shifted function. In case of real signal patterns, the multiplication process of $f(x+x'/2)f(x-x'/2)$ is produced at the object plane by passing the light twice through the transparency object, once with an x' inversion due to reflection caused by a roof-top prism. The next step is Fourier transform performed by a combination of spherical and cylindrical lenses, which serves as a Fourier transformation and an image formation in the x' and x directions. Hence, the WDF appears as functions of space and the spatial frequency in the x and x' directions, respectively, at the observation plane.

A 1-D specklegram of the 1-D displacement of a diffusely transmitting object is produced through two-step processes. In the first step, we produce an image of the 2-D specklegram of the 1-D displacement by slightly shifting the object between two exposures on a single film. The second step produces the 1-D specklegram of the 1-D displacement by using the 2-D specklegram of the 1-D displacement as an object for the optical system shown in Fig.2. An image of the object is produced at the image plane by a combination of spherical and cylindrical lenses, which serves as a Fourier transformation in the y direction and an image formation in the x direction. Hence, we convert pairs of 2-D speckles, existing in a certain position of the specklegram, to pairs of 1-D speckles. Since the generated 1-D speckles are recorded on the negative film, we use a slide copying adapter to transform the pattern into the positive film. The resultant 1-D specklegram of the ground glass with zero displacement is shown Fig.3.

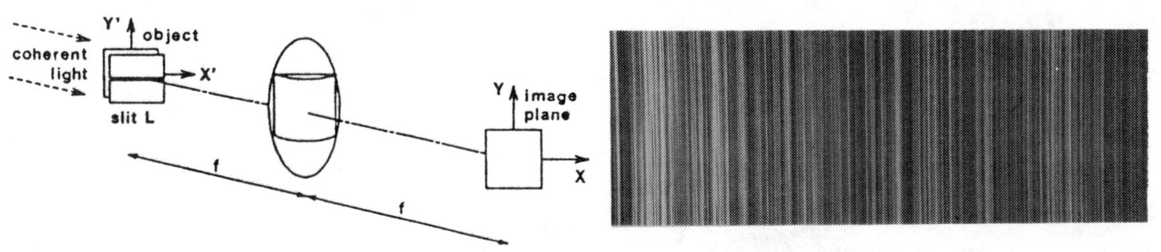

Fig.2. Optical systems for producing the 1-D speckle pattern.

Fig.3. Experimental result of the 1-D specklegram of a ground glass with zero displacement.

4. RESULTS AND DISCUSSION

Using the optical system of Fig.1, we produced the local spectra of both the 1-D simulated specklegrams of uniform and non-uniform (increasing) displacements and the experimental specklegrams of ground glasses and a rubber plate which are slightly shifted and stretched, respectively, between their exposures.

4.1. WDF production of simulated specklegrams

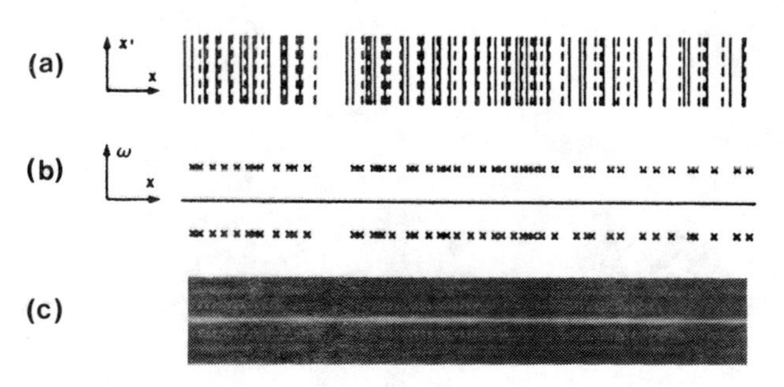

Fig.4. Simulated specklegram (a) of the uniform displacement, and calculated (b) and experimental (c) results of the local spectra $W(x,\omega)$.

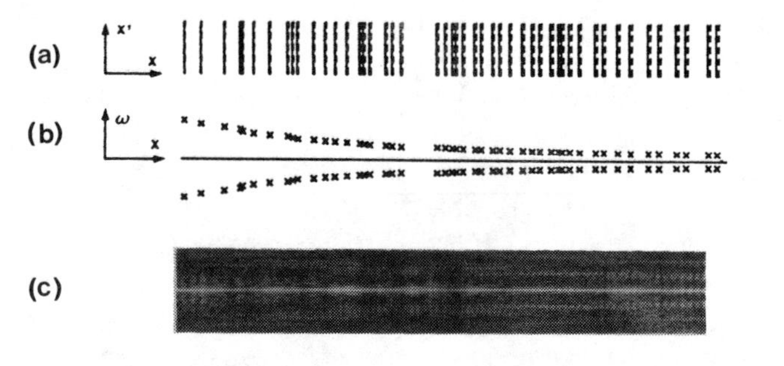

Fig.5. Simulated specklegram of the linearly increasing displacement (a), calculated (b) and experimental (c) results of the local spectra $W(x,\omega)$.

The 1-D simulated specklegrams of uniform and linearly increasing displacements and their local spectra $W(x,\omega)$, produced by using a computer and from an experiment, are shown in Fig.4(a)-(c) and Fig.5(a)-(c). Solid and dashed lines represent the 1-D speckle patterns, before and after shifting, respectively. The spatial frequency is expressed in the ω direction, while the x direction corresponds to the spatial position. Both of the theoretical and the experimental results show that uniform and linearly increasing displacements cause interference fringes to be constant and to change nonlinearly with respect to their positions. We can also observe the second-order fringes as the second lines along the x axis. These results can be understood by considering the spacings of Young's interference fringes which is given by

$$d = \lambda f / l, \tag{8}$$

where λ, f, and l are the wavelength of the read-out laser beam, the focal length of lens, and the interval of two pinholes. It is inversely proportional to the interval of two pinholes. Hence, wider displacement causes the smaller spatial frequency.

4.2. WDF production of specklegrams of the laterally shifted object

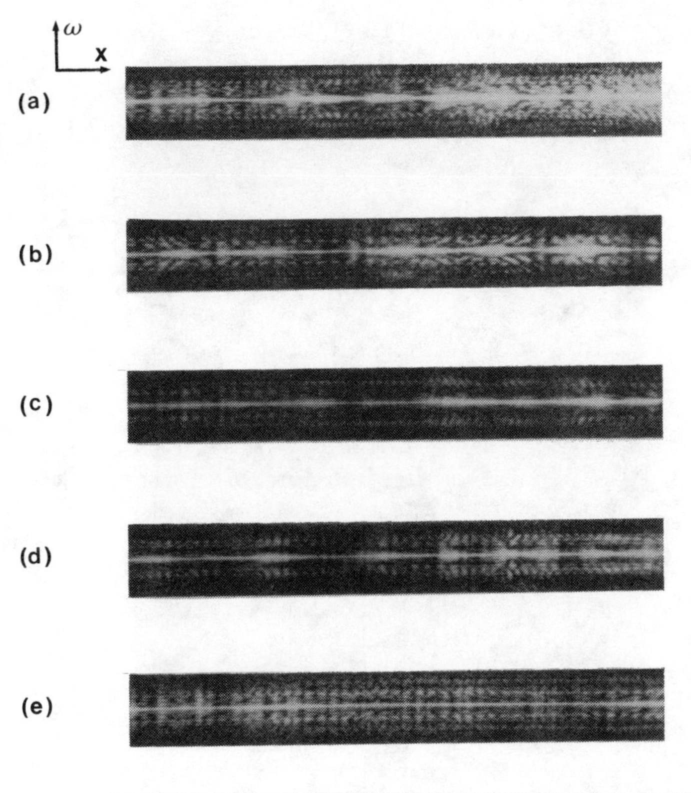

(f)

displacement (μm)	spacings of first-order fringes (d)	
	experimental (mm)	theoretical (mm)
60	0.9967 ± 0.0479	1.0546
80	0.7877 ± 0.076	0.791
100	0.5997 ± 0.0347	0.6328

Fig.6. Local spectra $W(x,\omega)$ for the speckles of 50μm size with (a) 0, (b) 40, (c) 60, (d) 80, (e) 100 μm displacements, and (f) experimental and theoretical magnitudes of spacing of the first-order fringes.

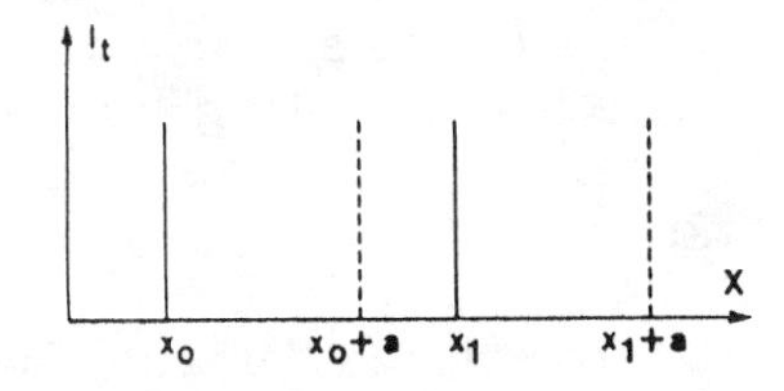

Fig.7. 1-D specklegram represented by the Dirac delta function.

Figures 6(a)-(e) show the local spectra $W(x,\omega)$ for the speckles of 50 μm size which are displaced within (a) 0, (b) 40, (c) 60, (d) 80, and (e) 100μm. It is clear from these figures that the interval of interference fringes observed in the ω direction changes in accordance with the range of displacements. When the displacements are zero or smaller than the size of speckles, interference fringes do not appear and cannot be observed as shown in Figs.6(a) and (b). As the displacement becomes larger than the speckle size, interference fringes appear in Figs. 6(c)-(e) and their spacings decrease proportionally with respect to the displacements. Especially in Fig.6(e), we can also observe the fringes of the second order. Comparison of the experimental and theoretical values of spacings of the first-order fringes is shown in Fig.6(f) and it shows that their values are nearly equal. Thus, the resultant local spectra $W(x,\omega)$ quantitatively agree with the theory.

Even though the local spectra $W(x,\omega)$ of the specklegram can be displayed by using the WDF, it is still not easy to differentiate the fringes from the noise. This problem, in fact, originally arises as a result of the process of the WDF itself, which will be shown below.

We assume that 1-D speckles can be expressed by Dirac delta functions which lead to the recorded intensity of a 1-D speckle pattern as

$$I = \sum_n \delta(x - x_n), \tag{9}$$

where x_n is the position of the n-th 1-D speckle. To simplify the discussion, consider that there are only two 1-D speckles. Then, the 1-D specklegram I_t can be expressed by

$$I_t = \delta(x-x_0) + \delta(x-x_0-a) + \delta(x-x_1) + \delta(x-x_1-a), \tag{10}$$

and is described in Fig.7, where a is a magnitude of the displacement, and the solid and dashed lines represent the 1-D speckles before and after shifting. In accordance with Fig.7, Eq.(4) can be depicted graphically in Fig.8(a) which takes place at the object plane of Fig.1. Substitution of I_t of Eq.(10) into f of Eqs.(3) and (4) leads to a result of the local spectra given by

$$W(x,\omega)= \delta(x-x_0,\omega) + \delta(x-x_0+a,\omega) + \delta(x-x_1,\omega) + \delta(x-x_1+a,\omega)$$
$$+ 2\delta\{x-(x_0+x_1)/2,\omega\}\cos\omega(x_1-x_0) + 2\delta\{x-(x_0-x_1)/2-a,\omega\}\cos\omega(x_1-x_0)$$
$$+ 2\delta\{x-(x_0+x_1+a)/2,\omega\}\cos\omega(x_0-x_1+a)$$
$$+ 2\delta\{x-(x_0+x_1+a)/2,\omega\}\cos\omega(x_1-x_0+a)$$
$$+ 2\delta(x-x_0-a/2,\omega)\cos\omega a + 2\delta(x-x_1-a/2,\omega)\cos\omega a, \quad (11)$$

which is shown in Fig.8(b). By taking Eq.(3) into consideration, the local spectra $W(x,\omega)$ of Fig.8(b) are generated by taking a Fourier transform of cross points of the multiplication of a 1-D specklegram in Fig.8(a). The numbers written in Figs.8(a) and (b) correspond to the successive number of terms in Eq.(11). The first four terms in Eq.(11), produced from the cross points of self-multiplication of 1-D speckles, contain constant spatial frequencies and can be interpreted as the background noise.

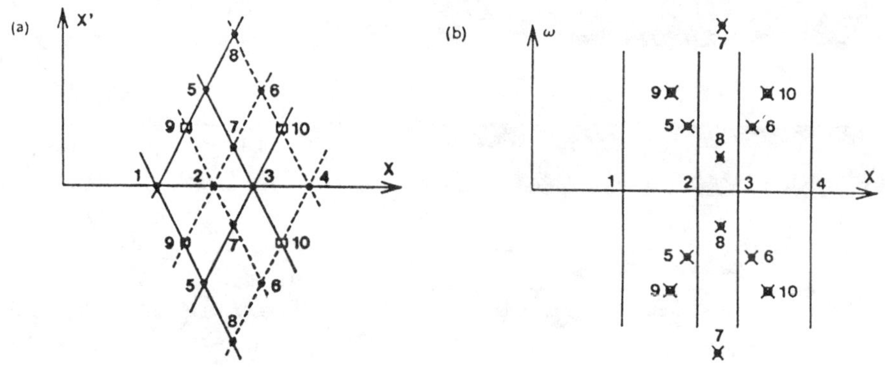

Fig.8. Multiplication of the 1-D specklegram at the object plane (a), Local spectra $W(x,\omega)$ at the image plane (b).

On the other hand, the information of the displacement, expressed by $\cos\omega a$ of the last two terms (ninth and tenth terms), emerges as a result of the pairs of cross points of cross-multiplication of correlated 1-D speckles. The fifth, sixth, eighth, and seventh terms have higher and lower spatial frequencies and come from the pairs of cross points of cross-multiplication of unshifted and shifted uncorrelated 1-D speckles. Both of the lower and higher spatial frequencies can also be considered as the noise and, together with the background noise, make poor the S/N ratio of the local spectra $W(x,\omega)$.

With knowledge of the cause of the noise, an appearance of the local spectra can be improved by eliminating at least the background noise, since the magnitude of the displacement is not known precisely in practical situations. In the above discussion, it was shown that the background noise came from the cross points of self-multiplication of 1-D speckles existing on the x axis. Under this consideration, these cross points can be prevented from being Fourier-transformed by using a window function of the transmittance in Fig.9(a) which is placed directly behind the object. The use of this window blocks the center where these cross points exist, and transmits the other part of the cross points. The improved local spectra $W(x,\omega)$ obtained from the windowed specklegrams are shown in Figs.10(a)-(c). Comparison of Figs.10(a)-(c) with Figs.4(c)-(e) shows that the fringes are sharpened and that the quality of the local spectra have been improved. However, the present method is limited by the aberrations of the lenses used in the optical system of Fig.2. This effect can be observed in Fig.3 where the 1-D speckles generated by the system do not have perfectly straight shapes as can be represented by the Dirac delta function. In addition, there is a limitation of the film, occuring especially during the transformation from the negative to the positive 1-D specklegram, which causes the two speckle patterns recorded to be not absolutely correlated. Although such limitations exist in the present method, both the qualitative and quantitative results, which will also be shown in the next discussion, show the feasibility of evaluating the displacement of objects using the WDF.

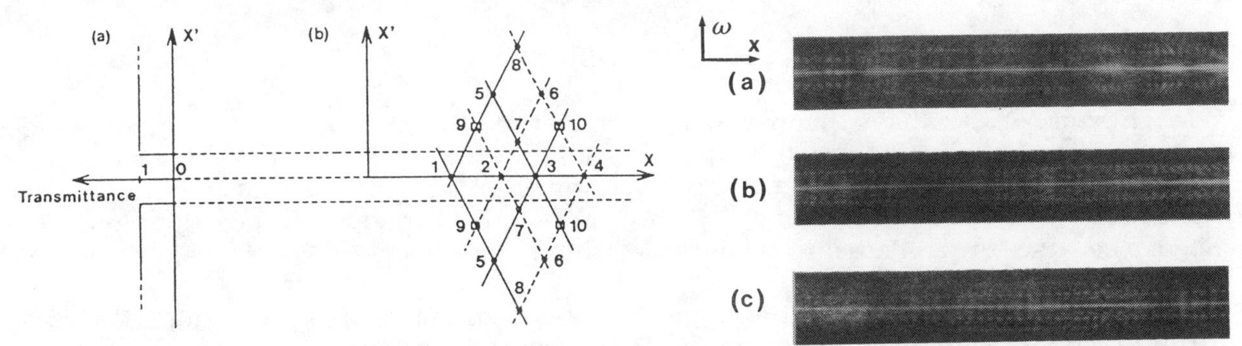

Fig.9. Transmittance of the window function (a) and its position with respect to the object (b).

Fig.10. Results of the local spectra $W(x,\omega)$ obtained from the windowed specklegram with (a) 60, (b) 80, and (c) 100 μm displacements.

4.3. WDF production of specklegram of the laterally stretched object

The same method used in Section 4.1 is also applied to the laterally stretched rubber plate shown in Fig.11(a). The result of Fig.11(b) shows that the spatial frequencies of the specklegram for the 120μm displacement change nonlinearly from the fixed end to the stretched end from the low to the high spatial frequencies, since the speckles of the stretched end have larger displacements than the other part of the rubber plate, especially the fixed end. In order to verify the quality of the local spectra, we compare it with the usual Young's interference fringes taken from the stretched, central, and fixed areas

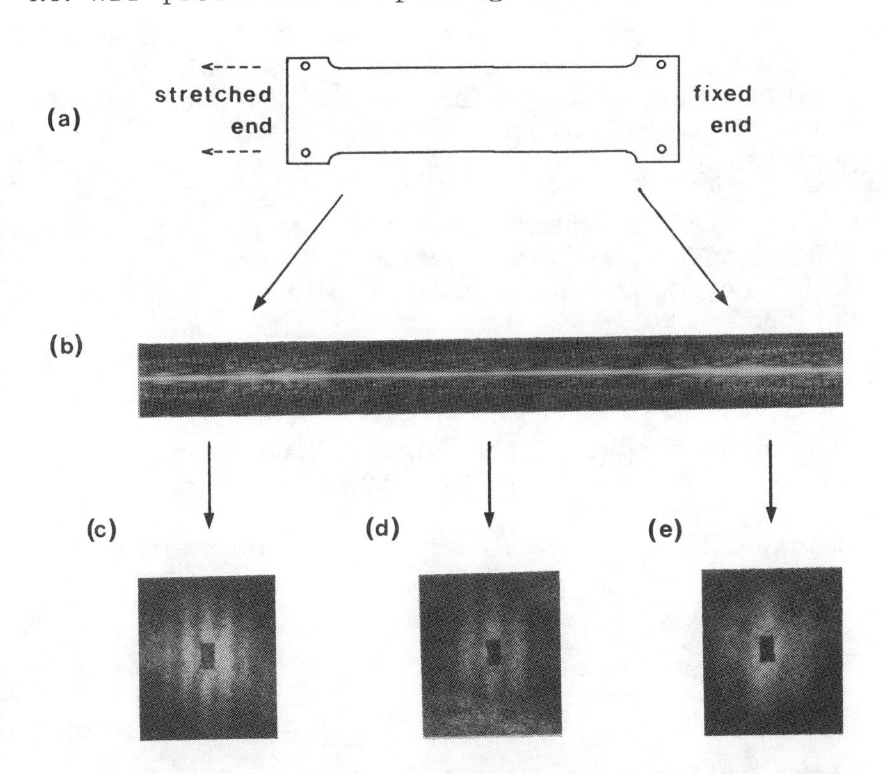

Fig.11. Laterally stretched rubber plate (a), its local spectra $W(x,\omega)$ of 120 μm displacement (b), and Young's fringes obtained by viewing from stretched (c), central (d), and fixed (e) areas of the same 2-D specklegram of the stretched rubber plate.

of the same specklegram of the rubber plate. The resultant Young's interference fringes are shown in Figs.11(c), (d), and (e). From these figures, it can be seen that, with respect to the positions, the spacings of the fringes are changing from the narrow to the wide spacings. Therefore, it can be concluded that the results of the local spectra obtained by using the WDF follow the Young's fringes. Since the spacing of the second-order fringes of Fig.10(c) are nearly equal to half the first order in Fig.10(b), the discontinuity of fringes of Fig.10(b) which appears near the stretched end may also be interpreted as the second-order fringe.

5. CONCLUSION

In this paper, we have proposed a new method to evaluate the local displacements of objects in speckle photography by using the WDF. The WDF has advantages over the usual Young's interference fringes in the point that it can display simultaneously the spatial frequency as a function of the spatial position and does not need the use of a Fourier transform which has to be taken regionally and repeatedly.

The experimental results, however, showed that the present method is not absolutely free of the noise and has a drawback which comes originally from the process of the 1-D WDF and also from the limitation of the optical system used to converte the 2-D to the 1-D speckles.

6. REFERENCES

1. C.P. Janse and J.M. Kaizer, "Time-Frequency Distributions of Loudspeakers: The Application of the Wigner Distribution," *J. Audio Eng. Soc.* 31, 198-223, 1983.
2. G. Cristobal, T. Bescos, J. Santamaria and J. Montes, "Wigner Distribution Representation of Digital Images," *Pattern Recognition Lett.* 5, 215-221, 1987.
3. M.J. Bastiaans, "The Wigner Distribution Function Applied to Optical Signals and Systems," *Optics Comm.* 25, 26-30, 1978.
4. R. Bamler and H. Glunder, "The Wigner Distribution Function of Two Dimensional Signals Coherent-Optical Generation Display," *Opt. Acta* 30, 1789-1803, 1983.
5. M. Corner and Y. Li, "Optical Generation of the Wigner Distribution of 2-D Real Signals," *Appl. Opt.* 24, 3825-3829, 1985.
6. T. Iwai, A.K. Gupta and T. Asakura, "Simultaneous Optical Production of the Sectional Wigner Distribution Function for a Two-Dimensional Object," *Optics Comm.* 58, 15-19, 1986.
7. T.A.C.M. Claasen and W.F.G. Mecklenbrauker, "The Wigner Distribution-A Tool for Time Frequency Analysis," *Philips J. Res.*, Part I 35, 217-250, 1980; Part II 35, 276-300, 1980; Part III 35, 372-389, 1980.
8. K.H. Brenner and A.W. Lohmann, "The Wigner Distribution Function and its Optical Production," *Optics Comm.* 32, 32-38, 1982.

SESSION 3

Surface Metrology and Testing

Chair
Craig W. Johnson
Plummer Optics Pte. Ltd. (Singapore)

Non contact optical microtopography

Manuel F.M. Costa , Jose B. Almeida

University of Minho, Physics Departement
Tel.(351 53) 612234, Telex 32135 rtumin P, Fax (351 53) 612368
P- 4719 BRAGA CODEX , PORTUGAL

ABSTRACT

This paper proceed with a revision of the basic principles involved in a method of optical microtopography that we are developing, presenting recent improvements. It is shown that a collimated light beam with an oblique incidence on a surface can be used to assess its distance from a reference plane if the bright spot produced on the surface is imaged onto an array of detectors which tracks its lateral displacement. The light beam is swept over the surface, so that large areas can be scanned.

An account is given of some pratical applications of the system. The system has been used with sucess to measure surface roughness of machined surfaces, thickness measure of silver and cooper thin films, for the topographic inspection of the edge of silver films sputtered through different masks, for the surface inspection of polyehylene films and for non contact measurement of fabric thickness and relief mapping.

1. INTRODUCTION

The need to meet the ever increasing industrial demands has promoted the appearance of different surface evaluation systems. Among these, non contact optical systems have gained a well deserved attention due to their particularly good adaptation to specific applications.

Roughness measurements and microtopographic inspection of rough surfaces requiring measuring ranges from some microns to a few milimeters, for roughness measurements from a few to hundreds microns preserving a high lateral resolution, are subject of growing concern and several profilers and microtopographers based on triangulation, focus sensing, moiré, ..., have recently been reported.

Devoted in first instance to the determination of fabrics parameters, we have developed a non contact microtopography system that we describe briefly below, presenting new developments and an application to the evaluation of polimers surfaces.

2. PRINCIPLE

The method is based on a triangulation procedure illustrated in Fig.1. The surface under inspection will be swept, point per point, by an oblique light beam, which creates on it a bright spot whose position is tracked by a microscopic system with axis perpendicular to the surface plane.

Fig. 1.

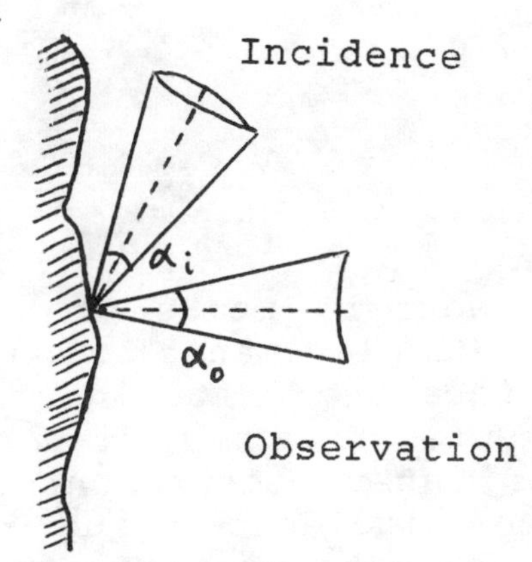

Fig. 2.

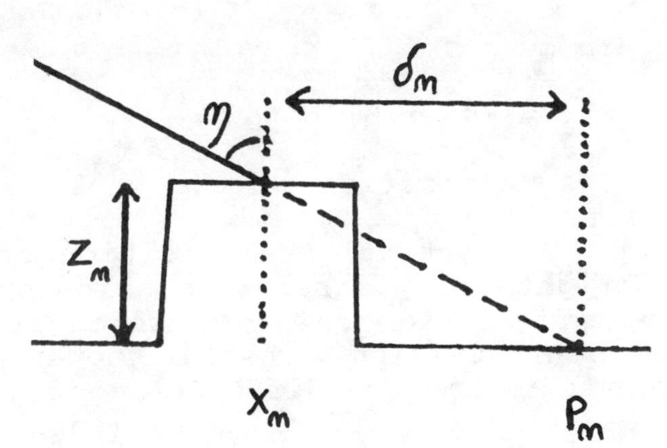

Fig. 1. and 2. The basic principles.

Fig. 2. illustrates the method: the intersection of the light beam with the surface creates a bright spot at a position *Xn*, the really sampled surface point, that corresponds to the *Rn* position on the swepping process that is the intersection of the direction of the light beam with the reference level. The lateral shift of the bright spot is proportional to the normal distance between the surface and the reference level:

$Zn = (\delta n/M) \operatorname{cotg} \eta$, where η is the angle of incidence of the light beam and M is the magnification of the microscopic optics.

In actual use the light beam is stationary and the surface is moved laterally, by equal increments Δ in the direction defined by the intersection of the plane of incidence and the plane of the surface. Usually η and Δ are known values and only the lateral shifts of the bright spot, δn, must be measured and recorded for later processing.

Assuming that the position *Po* is the origin of the coordinate system we get:

$Pn = n\Delta$;

$Xn = n\Delta - \delta n$;

$Zn = (\delta n/M) \operatorname{cotg} \eta$.

The profile of the surface, i. e. the line of intersection of the surface and the incidence plane, is the line linking all the points of coordinates:

$Xn = n\Delta - \delta n$,

$Zn = (\delta n/M) \operatorname{cotg} \eta$.

In order to build a 3D map of the surface, several profiles are taken separated by a distance σ, normal to the incidence plane and so the complete set of coordinates will be:

$$X_{n,m} = n\Delta - \delta_{n,m} \; ;$$

$$Y_{n,m} = m\,\sigma ;$$

$$Z_{n,m} = (\delta_{n,m}/M)\ \mathrm{cotg}\,\eta .$$

The calibration of the whole system is done by means of a smoth reference surface which is given precise displacements along its normal. At each vertical position of the surface the shift incurred by the bright spot is recorded allowing experimental determination of the conversion factor $\mathrm{cotg}\,\eta/M$ which is used for later measurements while the configuration rests unchanged (same incidence angle and magnification). For different configurations the conversion factors can be easily calculated or experimentally determined.

The depth resolution is limited, in essence, by the modifed Rayleigh limit. With a resolvable lateral shift:

$\delta_{min} = \lambda/\sin\alpha_o$; been $\sin\alpha_o$ the aperture of the reception microscopic optical system (Fig. 1.) and λ the wavelenght of the incident beam.

The depth resolution will be:

$$(\delta_z)_{min} = (\lambda/\sin\alpha_o)\ \mathrm{cotg}\,\eta$$

The measuring range will depend on the particular systems implementation and ranges from several tens of micron to some millimeters with depth resolution from some tenths of microns to several microns are possible. The area to be analized can be as large as desired with positioning resolutions in the micron range allowed by precise X,Y positioning stages.

3. SYSTEMS

Different implementations of this method are possible. We built a system[2] (Fig.3.) which is both simple and versatile, leading to the possibility of application to different situations: thickness measurement, roughness measurement and relief mapping of samples requiring different measuring ranges and resolutions.

The light source is an HeNe TEM00 laser at 632.8 nm and 1mw power. The laser is mounted on a rotational stage allowing incidence angle changes. The incidence optical system comprises a filter for optical power control and a lens system which focuses the beam onto a spot of reduced dimensions.

The reception optical system, which tracks the position of the bright spot consists of an interchangeable microscope objective (x10, x5 and x20) placed in front of a video camera that images the spot onto a light sensing array. As video camera we used a CCIR standard (Saticon tube) and CCD camera.

Fig. 3.

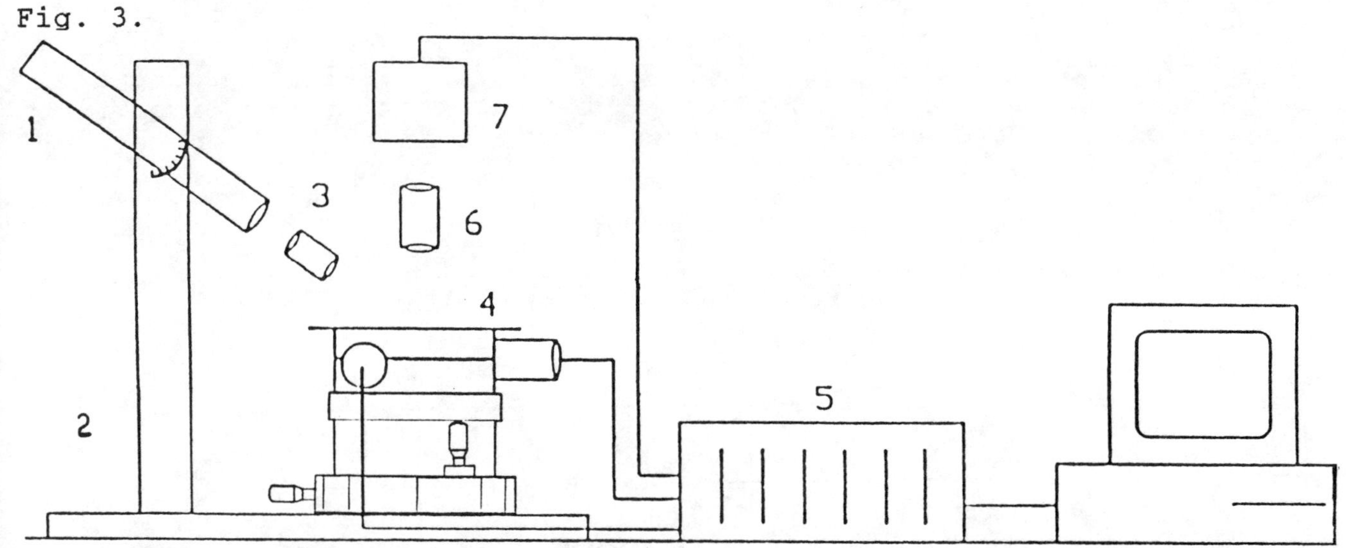

Fig. 3. The system: 1- HeNe laser; 2- Laser supporting structure; 3- Incidence optical system; 4- Sample suport and positioning setup; 5- Sample positioning and data acquisition control system; 6- Reception optical system; 7- Camera.

We chose to make the scanning of the sample surface by the light beam moving the sample under a stationary light beam. This solution poses less alignement and other optical problems than the reverse one, and larger areas can be scanned with high positioning acuracy and repetibility, at high speeds. The sample is placed on a reference surface provided with X,Y motion by two step motors alowing sampling of points on a rectangular array separated by distances down to 1 um. A rotational stage is also used to allow easy change to suplementary incidence in order to resolve shaded areas (usualy high incidence angles are used, depending on the surface caracteristics, to get higher conversion factors)[1] , and, a vertical movement precision stage is used for the calibration procedure.

A personal computer is used to control the "sample positionning and data aquisition control system". The data is later processed for rougness calculations or relief map plots.

4 APPLICATIONS

As we notice in the introduction the system has been developed to be used in thickness measurements and topographic inspection of fabric samples. Several kinds of fabrics have been tested and, as far as we know, for the first time without deforming the samples in the measuring processus[2,3].

Proving system's flexibility other kinds of samples have been already tested with promising results: roughness measurements of paper samples; thickness measurements of silver and cooper sputtered films[2,5]; topographic inspection of the borders of thin silver films produced by sputtering, with a planar magnetron source, through different masks, for which we prepared the system to give the best resolution of a few tenths of micron[2]; roughness measurements of machined surfaces; and microtopographic inspection of poliethilene films. As an example we proceed with the presentation of a table (Fig. 4.) showing the results of roughness measurements of some standards of surface rugosity, and a profile of a poliethilene film (Fig. 5.). The film as been produced in a extrusion machine with a old damaged vein that caused important imperfections on the film. The inspection of the film has been made to evaluate the extent of the vein damages.

Fig. 4.

STANDARDS OF SURFACE RUGOSITY

(RUGOTEST Nº 1 FROM ROCH, LTD)

	1	2	3	4	5	6	7	8	9
	SIDE MILLING			FACE MILLING			PLANING / TURNING		
SPECIFIED ROUGNESS µm(r.m.s.)	12.5 +/- 1.5	6.3 +/- 0.8	3.2 +/- 0.4	12.5 +/- 1.5	6.3 +/- 0.8	3.2 +/-0.4	12.5 +/-0.5	6.3 +/-0.3	3.2 +/-0.1
OUR MEASURE (r.m.s.) µm +/- 0.3 µm (20º, 20X)	12.9	6.5	3.2	13.9	7.1	3.7	12.6	6.5	3.4

Fig. 4. Roughness measures of machined surfaces.

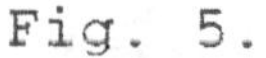
Fig. 5.

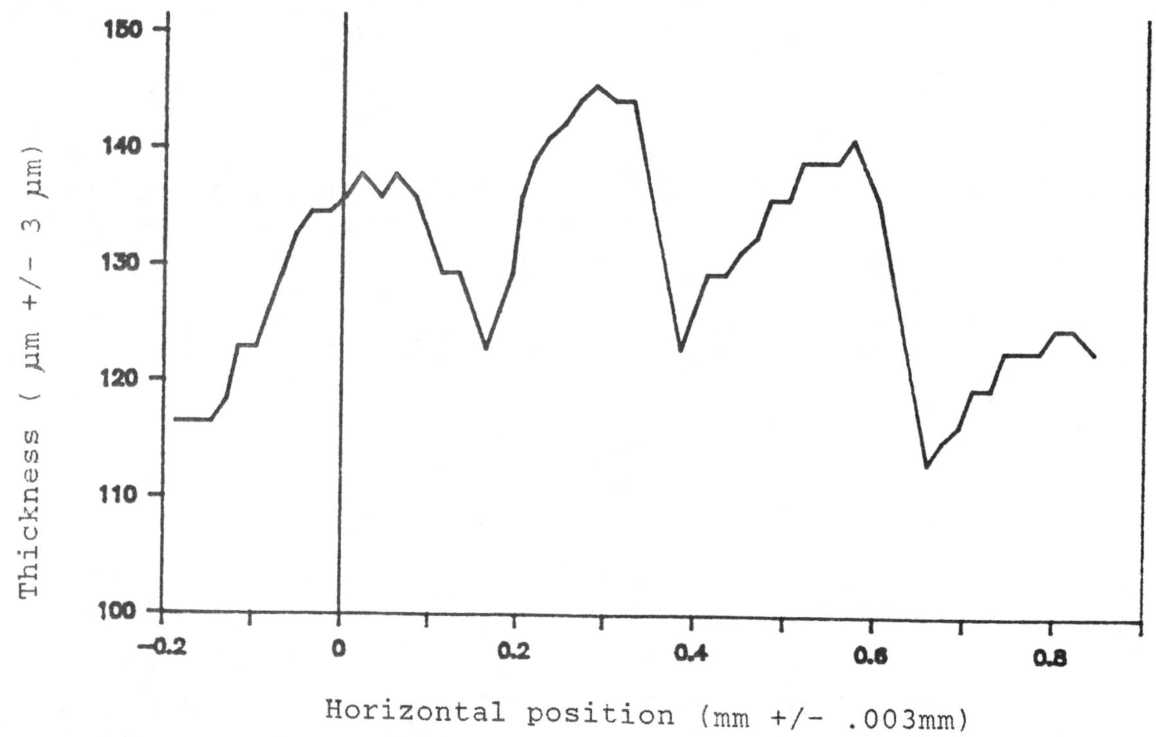

Fig. 5. Poliethilene film profile.

5. CONCLUSION

In the attempt of increasing the measuring range and resolution of the system, several improvements have been studied and are being implemented:

A diffraction-free *Jo Bessel*[7] beam generation setup has been studied as a possible solution to the depth of focus problems in the incidence optical system. Attempts will be made to produce beams with "radius" less than 5 µm;

In order to increase the heigth measuring range, the system will also be provided with a "macroscope" of high, adjustable, depht of focus;

Changes in the data acquisition system and in the processing procedures allow a substantial improvement in the system's speed.

6. REFERENCES

1. Manuel F.M. Costa, J.B. Almeida; "Surface relief mapping", SPIE vol. 1010, pag. 193/199, Hamburg 1988.
2. Manuel F.M. Costa, J.B. Almeida; "Surface microtopography of thin silver films", SPIE vol. 1332, San Diego 1990.
3. Manuel F.M. Costa, J.B. Almeida; "Mapping of textile surfaces relief", SPIE vol. 952, Porto 1988.
4. Manuel F.M. Costa, J.B. Almeida; "A system of non contact microtopography", Review of Progress in QNDE, La Jolla 1990.
5. R. Barral et al; "Construction and caracterization of a thin film vacuum coating system", Vacuum; vol. 39, pag. 843; 1988
6. G. Hausler; SPIE vol. 1319, Garmish-Partenckirchen 1990.
7. J. Durnin et al; "Diffraction-free Beams", Physical Review Letters, vol. 58, 15, pag. 1499, 1987.

Laser Scan Microscope and Infrared Laser Scan Microscope - two important Tools for Device Testing

Eberhard Ziegler

ICT GMBH, 8011 Heimstetten bei München, Germany

ABSTRACT

The optical beam induced current (OBIC) produced in devices by a laser scan microscope (LSM) is used to localize hot spots, leakage currents, electrostatic discharge defects and weak points. The LSM also allows photoluminescence measurements with high spatial and energy resolution. Using the infrared laser scan microscope (IR LSM), defects in the metallization and latch-up sensitive region could be detected from the back of the device.

1. INTRODUCTION

A scanning focussed laser beam produces in a LSM not only a contrast rich surface image with high resolution [1-3] but also generates many interaction effects between the laser beam and the specimen [4].

Both the surface image and the interaction effects allow the testing of samples in the fields of material science, biology, medicine and semiconductor physics.

The Infrared Laser Scan Microscope (IR LSM) makes use of infrared laser light[5]. Moreover, the optics must be corrected for infrared light and coatings for infrared are preferable. In this paper a combined LSM-IR LSM instrument and some new applications are presented.

2. SET UP OF THE COMBINED LSM - IR LSM

Fig.1 shows the system diagram of the equipment. It consists of three parts: the optical system (on the left hand side of Fig. 1), the electronic system (middle) and the image processing and frame store system (on the right).

The optical system may be connected to light microscopes with special coatings, but also some commercial light microscopes can be used. The red He-Ne-Laser (633 nm) is included in the optical system, the blue argon ion laser (488 nm) and the infrared He-Ne-Laser (1,152 µm) are positioned outside the optical system and connected to it by two single mode glass fibers. The core diameter for the infrared is about 9 µm, for the blue 4 µm. The transmission efficiency of the cables, including the coupling optics to the laser and to the optical system, is 50 % for 488 nm and 70 % for 1,152 µm.

In the optical system, the laser beam can, if required, be pulsed by an acousto-optical modulator. A system of deflecting mirrors, controlled by a digital scan generator, is used to position the laser beams in x and y directions.

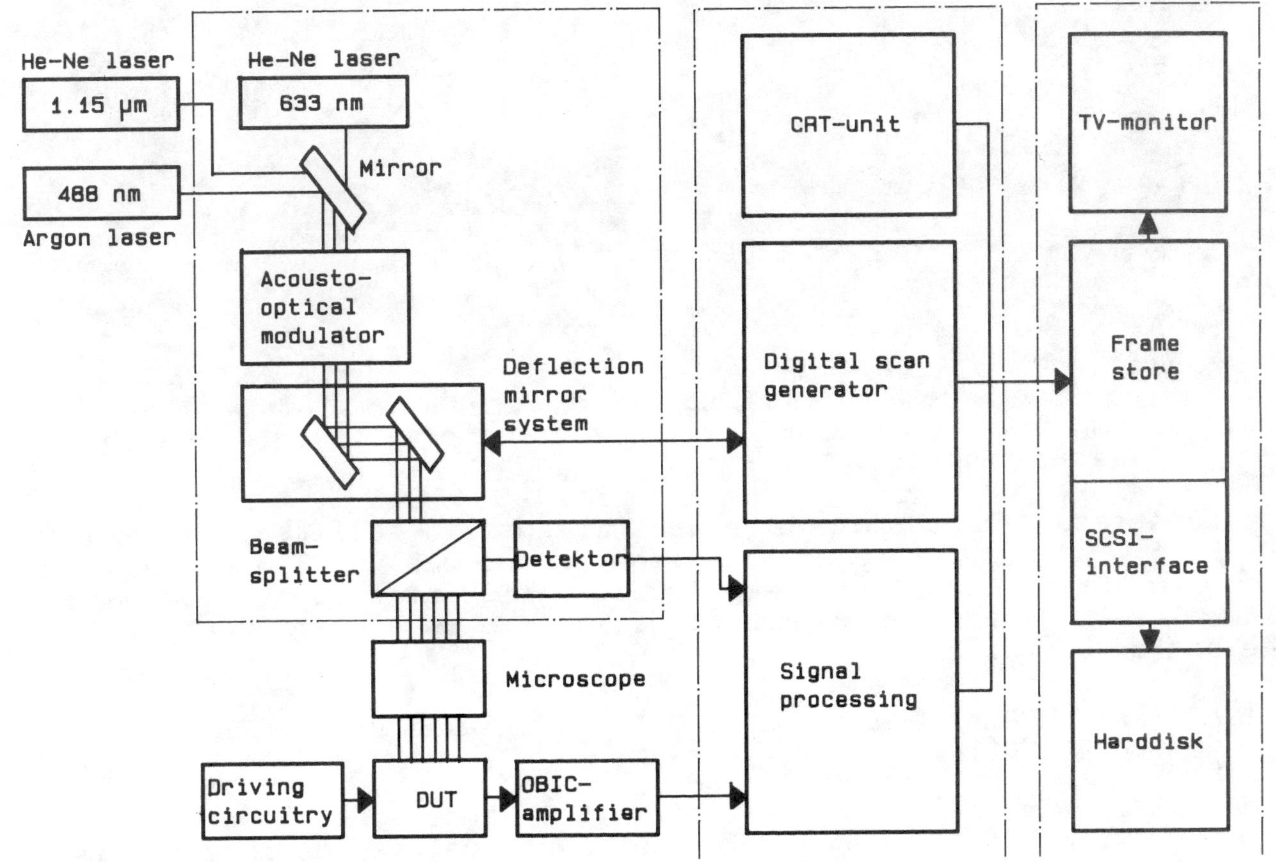

Fig.1. Block diagram of a combined Laser Scan Microscope and Infrared Laser Scan Microscope.

The light microscope focusses the scanning beam onto the sample. The reflected light is detected by a silicon- or germanium photodiode. Absorbed photons generate effects such as the optical beam induced current (OBIC), or photoluminescence.
The signals can be directly imaged on a slow-scan monitor, or stored in a frame store. The frame store not only converts the slow-scan image into a TV-image, but also contains several image processing routines. The SCSI interface of the frame store allows the transfer of images to the hard disk and the read-out of images from the hard disk. Usually we employ a 80 Mbyte hard disk.

3. APPLICATIONS OF LSM IN DEVICE TESTING USING THE OBIC EFFECT

A laser beam generates electron-hole pairs in semiconductor material if the energy of its photons is larger than the band gap of the semiconductor. If an internal electric field exists or an external electric field is applied, the carriers are separated and generate a current, known as the optical beam induced current. The penetration depth of the laser beam in the semiconductor depends on its wavelength. Therefore it is important to work with different laser wavelengths, if one needs information from different depths in the specimen. Moreover, the lateral resolution of the reflected light image depends on the wavelength.

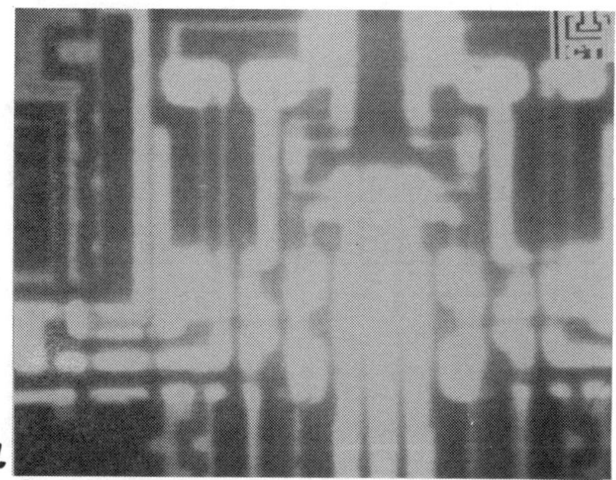
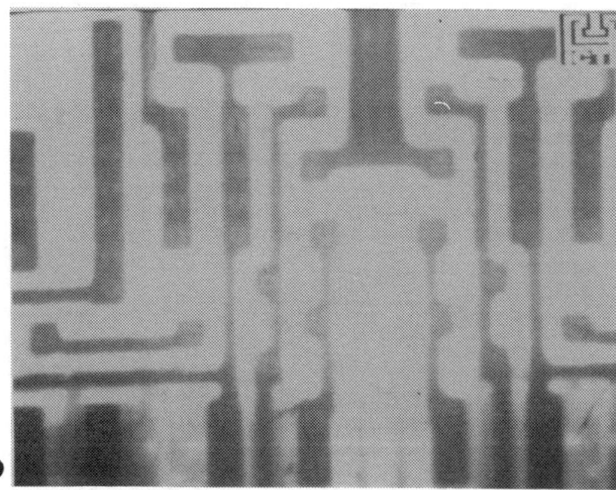

Fig.2. Comparison of the OBIC signals using blue (a) and red (b) lasers, respectively.

The penetration depth for blue 488 nm light in silicon, for example, is about 1 µm and for red 633 nm light about 2,5 µm. Fig. 2 shows that the OBIC images of a device are quite different using the red and the blue laser, respectively. The electric fields at different depths of the device are different.

3.1. Localization of leakage currents, hot spots and electrostatic discharge defects (ESD).

The OBIC contrast contains information about all these kinds of defects, which cause a change in the electrical fields in the interior of the device. Nevertheless, it is sometimes difficult to decide if the OBIC contrast shows a defect or not, especially in high integrated devices.
The user must be familiar with the interpretation of the OBIC contrast as well as with the design of the device. A simple method to localize these kinds of defects is a "GOLDEN DEVICE TEST".

That means, the OBIC signal of the device under test is subtracted from the OBIC signal of a correctly functioning device (GOLDEN DEVICE) of the same type. In order to get an exact superposition of the same region of both samples, a crosshair is used to choose the same reference point in each image. The subtraction is performed in the frame store. Any dissimilarity in the OBIC images can be seen in a bright or colored spot in the difference image. This point or region points to a defect in the device under test.

Fig. 3 illustrates how it was possible to localize a hot spot in an IC by comparison with a golden device. Fig. 3 a shows the surface image of the region of interest taken with the reflected laser light. Here, the red He - Ne laser with the wavelength 633 nm, was used. Fig. 3b and 3c show OBIC images of the golden device and the device under test, respectively. Without a reference one would probably not conclude from the image in Fig. 3c that a defect is present. By comparing 3b with 3c one easily recognizes a difference between the OBIC images of the tested device and those of the golden device. Fig. 3d depicts this difference and thus reveals the defect in the device under test. With the help of liquid crystal thermography, we could confirm that the defect shown in Fig. 3d is a hot spot.For the localization of leakage currents and ESD-defects we use the same procedure.

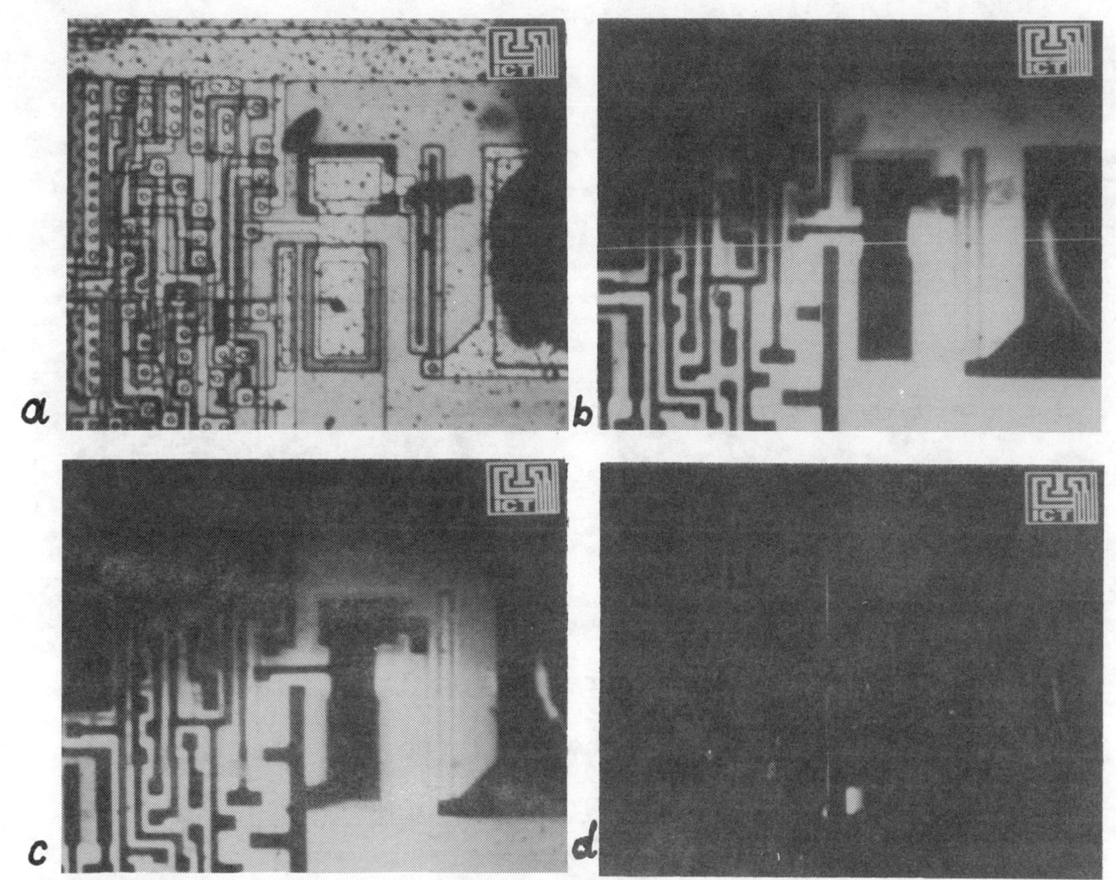

Fig.3. Localization of a hot spot using a golden device test. (a=reflected light image, b=OBIC image of the reference IC, c=OBIC image of the device under test, d=difference of the OBIC images).

3.2.Localization of weak points

During this test, a correctly functioning integrated circuit is operated under extreme conditions of voltage, frequency or temperature, respectively. If the laser beam strikes an area which is already close to failure, it disturbs the functioning of the device. The variation of the output signal, as a result of the laser beam influence, is used to locate the weak points in the IC.

An example of such a test is given in Fig. 4. Here, a frequency divider IC for wrist watches was investigated (working conditions: power supply 1,5 V; input signal 32 kHz, output signal 1024 Hz).
The left half of the photograph shows the reflected light image and the right half the output signal and its variation. The regular pattern represents the superposition of the normal output signal of 1024 Hz with the scan frequency. The dark and the bright regions signify a loss of the output signal and reveal the weak points.

Other more well known applications of the OBIC for device testing are the localization of the latch-up effect in CMOS-devices[6,7], the detection of logic states in CMOS devices[8], the localization of defect memory cells[4] and the detection of soft errors[4,9]. These applications are not discussed in this paper.

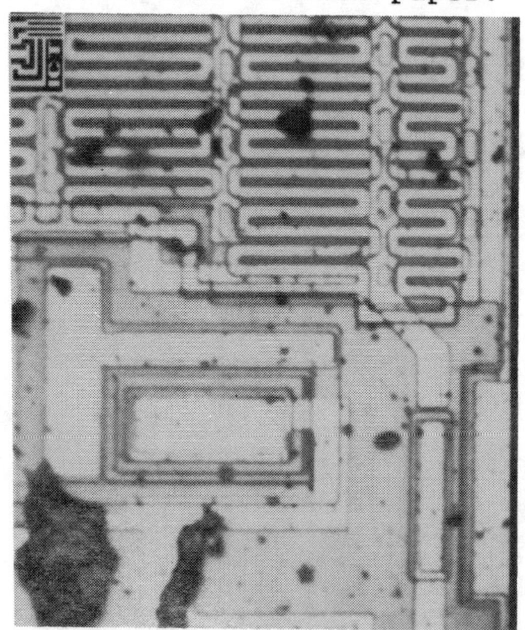

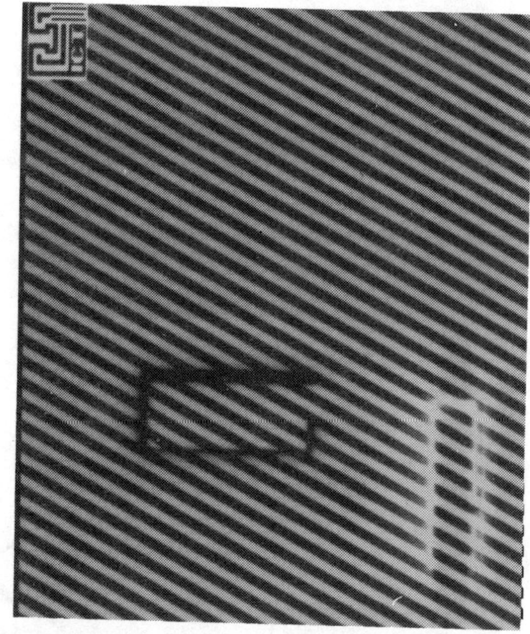

Fig. 4. Detection of weak points, reflected light image (left) and the output signal (right). The dark and bright regions reveal a change in the output signal and thereby the weak points.

4.PHOTOLUMINESCENCE WITH THE HELP OF THE LASER SCAN MICROSCOPE

By making small modifications to the LSM shown in Fig.1, one obtains a photoluminescence equipment with high spatial resolution and high energy resolution. For the excitation of the photoluminescence an argon ion laser (488 nm) with higher power (1≈W) is used. This laser radiation is transmitted with a single mode glass fiber into the optical system. The luminescence light and the reflected light is transmitted from the light microscope to a tunable monochromator with the help of a glass fiber bundle. At the exit of the monochromator a photomultiplier detects the luminescence intensity as a function of the wavelength. All other parts of the LSM are the same, as shown in Fig. 1.

With this equipment one can measure the luminescence intensity in relation to the wavelength, at a fixed point, or the luminescence intensity in relation to the location, at a fixed wavelength[10].
We have used luminescence intensity measurements for the test of GaAs/GaAlAs heterostructures.

5.APPLICATIONS OF AN INFRARED LASER SCAN MICROSCOPE

With modern high density integrated circuits with multiple layers of aluminium tests become difficult with the LSM because of the multiple reflecting metal layers. A solution would be to test devices from the back, through the silicon. We have developed such a method based on an IR LSM. The wavelength of the He - Ne - Laser (1,152 µm) used in the IR LSM corresponds exactly to the energy gap of silicon at room temperature (1.07 eV), so that silicon begins to become transparent. However, the energy of the photons of this laser is just great enough to produce electron-hole pairs. Thus one can obtain not only the reflected light image, but also the OBIC signal from the back of the integrated circuit through the silicon.

The uniqueness of this infrared laser make it suitable for some specific applications:

- Investigations are possible at arbitrary depths in an integrated circuit by moving the focal plane.
- Mechanical and electrical defects under the metallization can be imaged and localized.

In comparison with other methods, the IR LSM has the following advantages:

Infrared laser scan microscopy is nondestructive. The energy of the infrared photons (about 1 eV) is negligible in comparison with X-rays or electron beams and, in contrast to electrons, the photons have no charge.

The main advantage of IR LSM is in the generation of the OBIC from the back of the device, which is not possible with conventional infrared microscopy.

The only disadvantage of the IR LSM for device testing is the required preparation of the back surface. It is important that the infrared beam strikes a clean silicon surface.

5.1.Localization of defective contacts

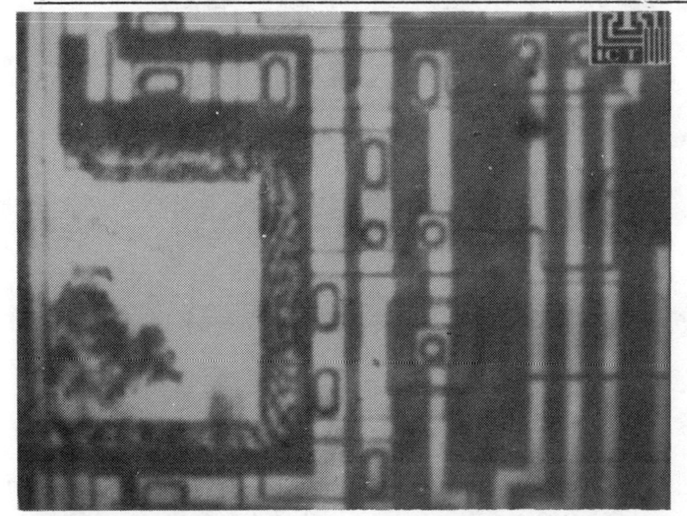

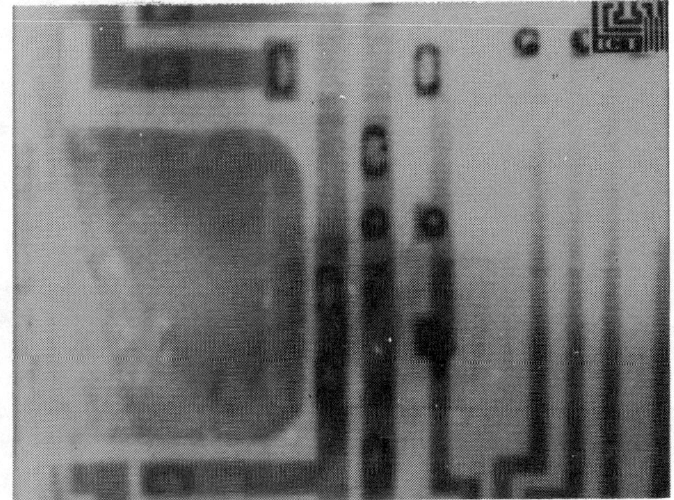

Fig.5 Defect in metallization. Reflected light image (left) and OBIC image (right) of the same region from the back of the device.

Fig. 5 exhibits the infrared reflected light image (5a) and the infrared OBIC image (5b) of the same area of a device from the back. The contact is electrically operational, although one can see some defects in the reflected light image and in the OBIC image.

The bright OBIC contrast (at top right of the OBIC image), is caused by a p-well. The laser beam strikes the p-n junction between p-well and substrate before reaching the metal lines and generates a large current. In the OBIC image most of the small contact holes show a high contrast around the hole.

5.2. Latch-up investigations from the back

In CMOS circuits, there always exist a parasitic thyristor structure. such a structure is prone to latch-up, which can lead to the destruction of the circuit.

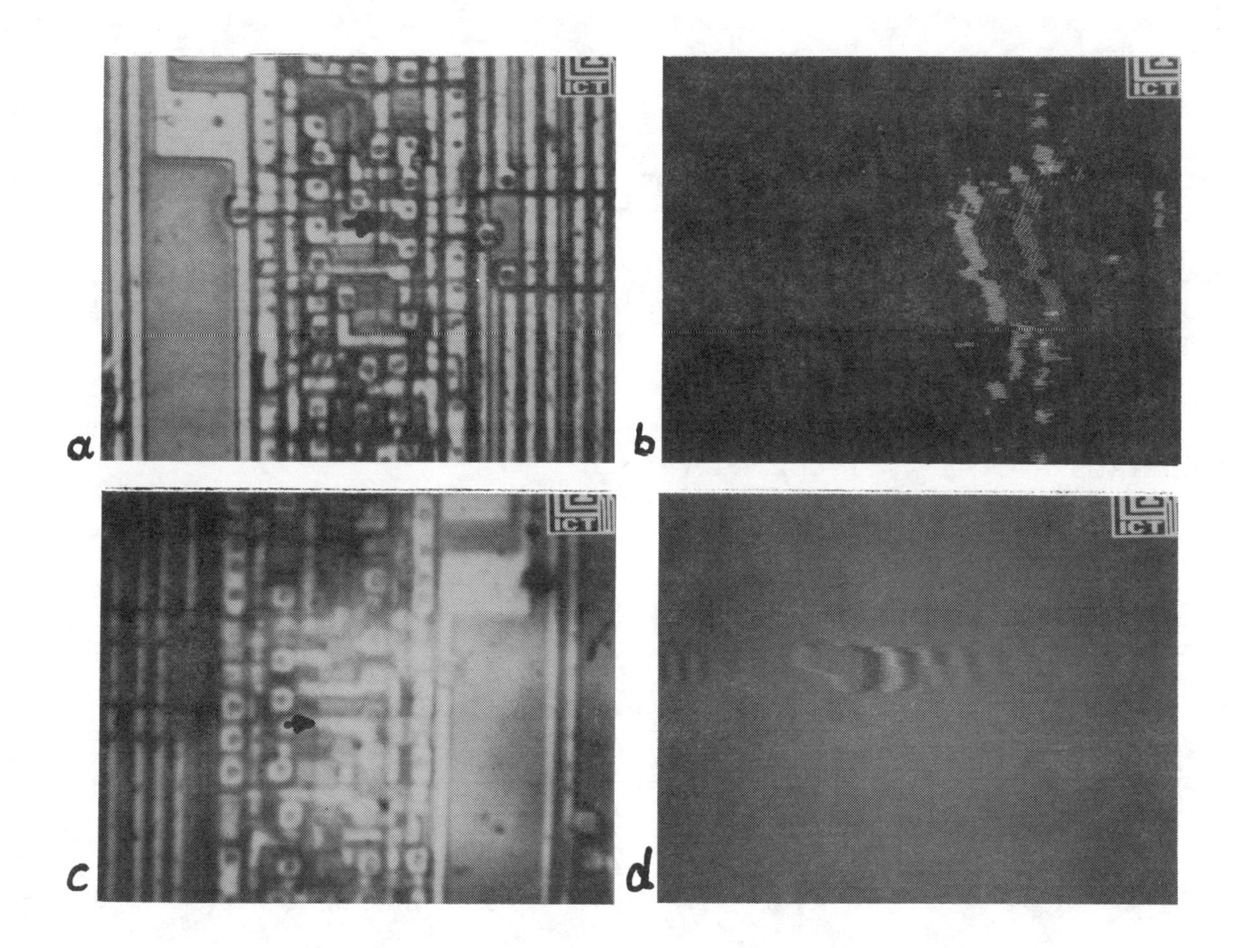

Fig.6. Localization of latch-up sensitive regions from the front and from the back of the device. a) Reflected light image from the front (633 nm laser), b) Latch-up current triggered by the 633 nm laser from the front c) Reflected light image from the back (1.152 µm laser) d) Latch-up current, triggered by the 1.152 µm laser from the back.

We have successfully localized the latch-up sensitive regions by irradiation with IR laser light from the back of the device. The aluminium layers on the front of the chip do not influence such investigations. During the measurements, the power supply of the device under test was pulsed, similar to tests from the front with visible laser radiation. The 1.152 µm infrared laser penetrates the silicon substrate from the back of the device and scans the underside of the metallization. The current produced by this laser can fire the thysistor-structure of the CMOS device.
Fig. 6. shows a latch-up sensitive region localized with red laser (633 nm) from the front and with the infrared laser from the back of the device. One can see that the latch-up sensitive region was localized with both methods at almost the same position, between well and substrate. The edges of the well are marked by arrows.

6. REFERENCES

1. C.J.R. Sheppard and A. Cloudhury; Optica Acta 24, 1051 - 1073 (1977)

2. C.J.R. Sheppard and T. Wilson; Optica Acta 25, 315 - 325 (1978)

3. G.J. Brakenhoff, P. Bloom and P. Barends; J. Microscopy 117, 219 - 232 (1979)

4. E. Ziegler, H.P. Feuerbaum; Microelectronic Engineering 7, 309 - 316 (1987)

5. E. Ziegler, H.P. Feuerbaum; SPIE Proc 1028, 226 - 230 (1988)

6. F.J. Henley, M.H. Chi, W.G. Oldham; IEEE/IRPS 21, 122 (1983)

7. R. Müller; Siemens Res.Repts 13,9 (1984)

8. F.J. Henley; Proceedings of the Custom Integrated Circuits Conference (CICC) p.181 May (1984)

9. E. Ziegler, H.P. Feuerbaum; ITG-Fachberichte 98, Großintegration 271 (1987)

10. E. Ziegler; SPIE Proc. 1139, 55-60 (1989)

PHOTON SCANNING TUNNELING MICROSCOPY

Jean Pierre GOUDONNET, L. SALOMON, F. DE FORNEL and G. CHABRIER
Université de Bourgogne, Laboratoire de Physique du Solide, Groupe Photoélectricité,
UA 785, Fac Sciences Mirande, BP 138, 21004 DIJON CEDEX - FRANCE

and

R.J. WARMACK and T.L. FERRELL
Health and Safety Research Division, Oak Ridge National Laboratory, Oak Ridge,
Tennessee 37831

ABSTRACT

The Photon Scanning Tunneling Microscope (PSTM) is the photon analogue of the electron Scanning Tunneling Microscope (STM). It uses the evanescent field due to the total internal reflection of a light beam in a Total Internal Reflection (TIR) prism. The sample, mounted on the base of the prism, modulates the evanescent field. A sharpened optical fiber probes this field, and the collected light is processed to generate an image of the topography and the chemical composition of the surface. We give, in this paper, a description of the microscope and discuss the influence of several parameters such as - polarization of light, angle of incidence, shape of the end of the fiber - on the resolution. Images of various samples - glass samples, teflon spheres - are presented.

I. INTRODUCTION

Photon Scanning Tunneling Microscopy (PSTM) belongs to the family of SXM microscopies [1,2] which exploit a distance dependent interaction for guidance of a probe tip and local information about this interaction. One of these new microscope - Scanning Tunneling Microscope [3] - demonstrated its capability to map atomic and molecular shapes, overcoming the fundamental limitation of any conventional microscope : diffraction obscures details smaller than about one half the wavelength of the radiation. The STM has been considerably described in the recent years and has imaged surface structure on a host of substance. However, the STM is restricted to imaging electrical conductors.

In 1988, at Oak Ridge National Laboratory, R.C. REDDICK, R.J. WARMACK and T.L. FERRELL [4] invented the PSTM which is a similar instrument to the STM but one which uses the tunneling of photons across the gap between the sample and a sharpened optical fiber probe tip. The PSTM uses a sample and a tip that are conductors for photons and is restricted to imaging transparent samples. If the sample is opaque it is possible to change the light source and switch to a wavelength at which the sample is not absorbing.

In chapter II a general description of the PSTM's operation is given and images of several samples is shown. A discussion of the optimization of the resolution of the microscope by acting on the important parameters is presented in chapter III. Application of PSTM - probing of optical waveguides, testing of optical surfaces are discussed in the last chapter.

II. PRINCIPLES AND OPERATION OF THE PSTM :

1) Imaging

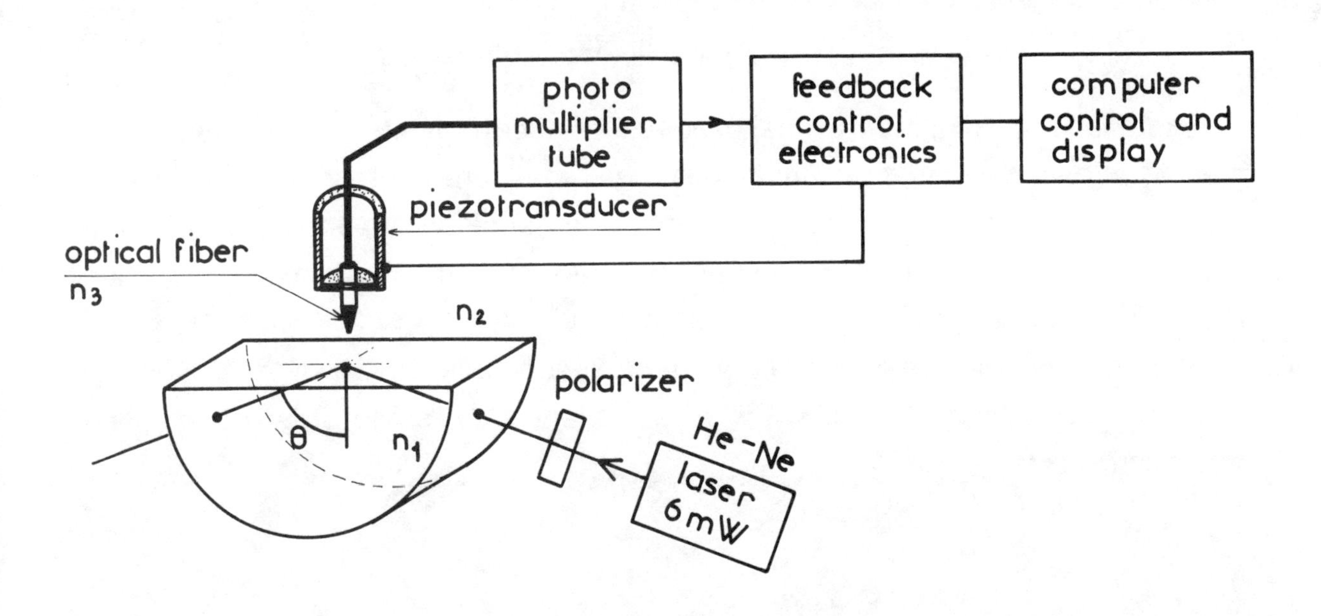

Figure 1 : Schematic of experimental set up.

An evanescent field is formed by total internal reflection at the interface between two semi-infinite media of different indexes of refraction n_1 et n_2 ($n_2 < n_1$). The light is totally reflected if the angle of incidence is larger than the critical angle $\theta_c = \sin^{-1} n_2/n_1$. This condition provides an exponentially decaying probability of photons tunneling from the sample surface to a fiber optic probe. The light is conducted by the fiber to a photomultiplier tube and converted to an electrical signal. The subsequent imaging process is identical to that of the STM.

The probe tip is mounted in a piezoelectric tube similar to those used in scanning tunneling microscopy (3). Voltages applied to the tube allow detailed control of the motion of the tip in three dimensions. Depending on the characteristics of the tube different scanning area are possible ranging from (1μm x 1μm) up to (100 μm x 100 μm) tangential to the sample.

The probe tip is formed by etching one end of a multimode graded-index optical fiber (50/125 CGE) in a solution of hydrofluoric acid.

The decay length of the exponentialy decreasing light intensity in medium n_2 can be calculated for p and s polarization by using the expression of the electric fields in the different media. These calculations show (5) that the decay length for a PSTM is a function of the shape of the tip, the angle of incidence θ, the polarization of the light, and the values of the three refractive indexes n_1, n_2, n_3. It is shown, also, that a decrease in decay length induces an increase in image resolution.

Figure 2 is an example of the exponential decay of the field intensity as a function of the increasing tip-sample distance, measured for several angles of incidence. In terms of decay lengths (or depth of penetration depending on different authors) it can be seen in Fig. 3 that a very small decay length (40 nm) can be reached by using a λ_2 = 2500 Å wavelength illumination and a near-grazing incidence. Since TIR may be observed for an enormous range of electromagnetic radiation, including X rays, we expect to endow the PSTM with resolution of the same order of scanning electron microscopy.

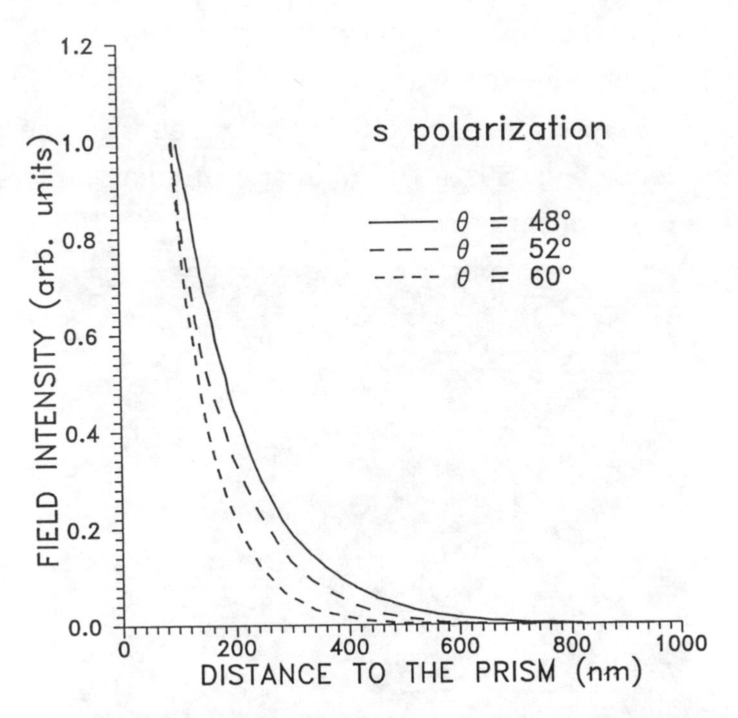

Figure 2 : Field intensity (Photomultiplier current) as a function of increasing distance d from the interface measured for s polarization and several angles of incidence at $\lambda = 6328$ Å.

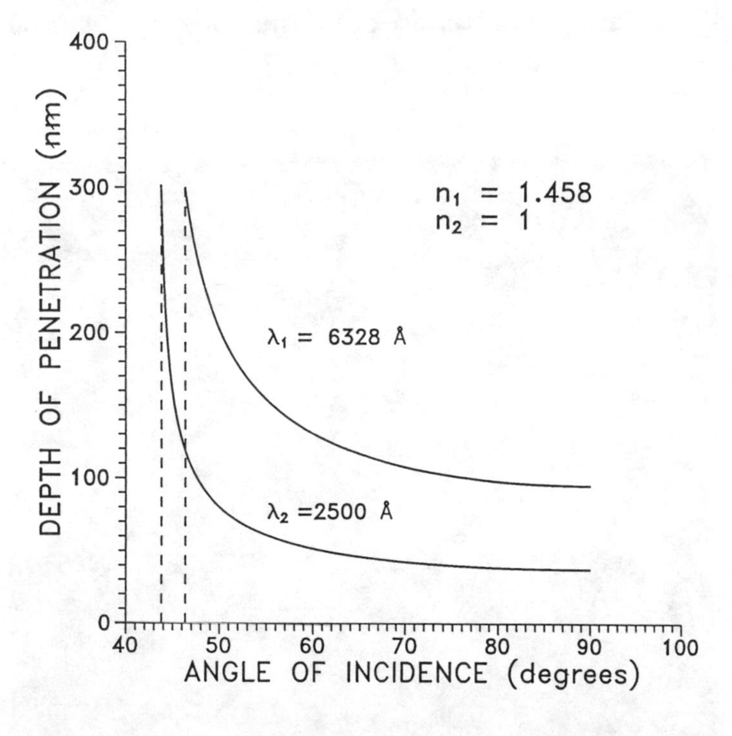

Figure 3 : Depth of penetration (or decay length) as a function of increasing angle of incidence for two wavelengths $\lambda_1 = 6328$ Å and $\lambda_2 = 2500$ Å. A glass-air interface has been considered.

PSTM images of two samples are shown in Fig. 4, and 5. Fig. 4 is the image of the surface of a quartz slide which was coupled to the prism base by an index-matching fluid. The subwavelength-sized roughness is seen with corrugations on the order of 10 - 20 nm deep. Fig. 5 is a PSTM image of teflon spheres (ø = 225 nm) deposited on the base of the prism.

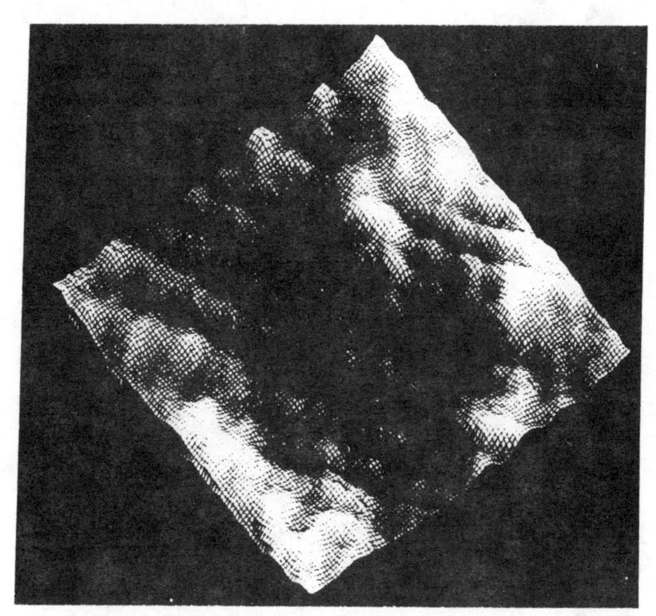

Figure 4 : Image of the surface of a quartz slide taken with a PSTM. The scan area is 3 x 3 μ^2.

Figure 5 : Image of teflon spheres on the base of the prism

III. RESOLUTION OF THE MICROSCOPE OPTIMIZATION

High spectral resolution being required, we have conducted a serie of theoretical and experimental studies in order to optimize the parameters responsible of the variations of the penetration depth d_p - shape of the tip, polarization, angle of incidence [5].

These studies show that for both s and p polarizations the decay length decreases as the angle of incidence increases. The smallest values of d_p are obtained for the largest values of the angle of incidence and for s polarization (Fig. 6). In the PSTM the shape of the tip could be thought to be critical for the coupling between the evanescent field and the third medium. As a matter of fact it is nct. The reason is that the strong dependence of energy flux on tip-to-surface distance makes only a fraction of the tip subtend any significant energy flux. This permits better tangential resolution than might be expected. The demonstration of this statement is easily made by assuming that the end of the fiber is a paraboloid (Fig. 7). It can be seen, that the variation of the intensity calculated for two models of paraboloid (a "needle" a = 4.10^7 and an "infinite plane" a = 0) are both similar to those of the experiments.

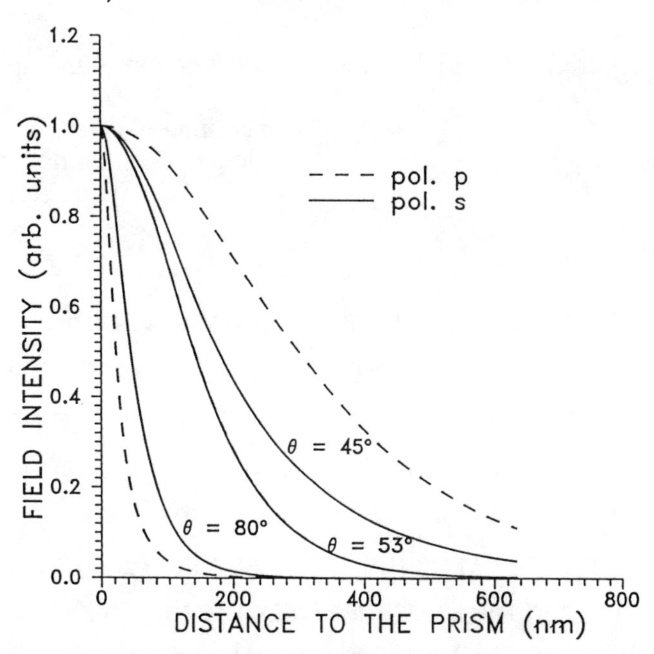

Figure 6 : Field intensity in medium 3 (optical fiber) calculated as a function of the distance to the prism and p (solid line) and s (dashed line) polarizations. For $\theta = \theta_l = 53°$ the intensity is the same for both polarizations. For the calculations the following values of the parameters have been used : $n_1 = n_3 = 1.458$, $n_2 = 1$, $\theta_c = 43°30$, $\theta_l = 53°$.

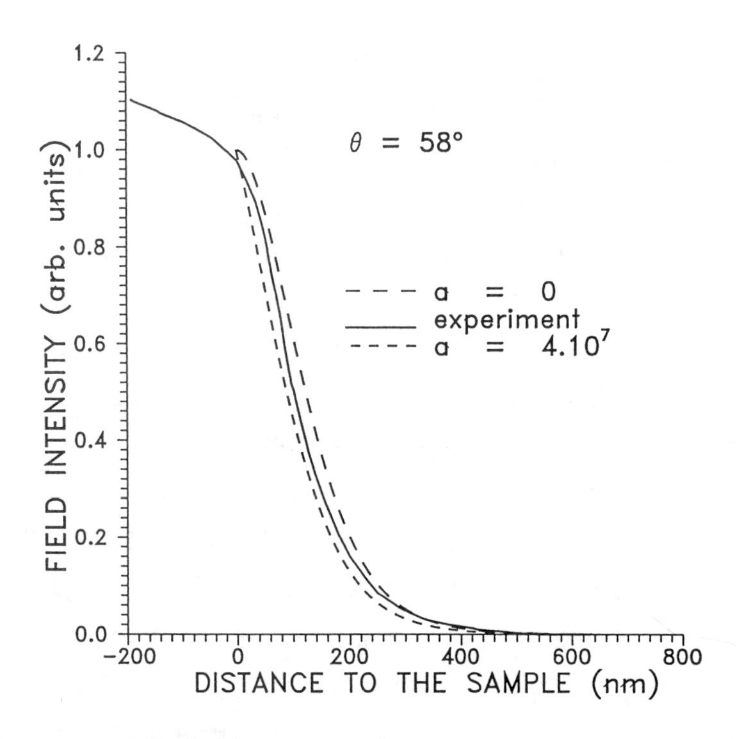

Figure 7 : Field intensity as a function of the tip-sample distance.

experiment
calculated with a flattered paraboloid model
calculated with a sharp paraboloid model.

If a sample to be studied has a spatially varying chemical composition, spectroscopy can be carried out simultaneously with the PSTM microscopy and a chemical map can be superimposed on the topology. The specific compound in a given region can be distinguished in the image by absorption or Raman spectroscopy. Results of these procedures will be published elsewhere (6).

The photoluminescence spectrum from a sample of ruby taken with a PSTM has been localy determined and the stress features present in the sample monitored in the region probed (7). The PSTM technique has been shown (8) to probe directly the evanescent field outside a planar and a channel waveguide. Surface scanning images in either constant height or constant intensity mode have been obtained and have the potential to reveal both local topographic and dielectric fluctuations.

CONCLUSION

PSTM offers many exciting possibilities for surface science research. A variety of applications are forthcoming, microlithography, spectroscopies. Biological work could benefit from high resolution chemical mapping of samples in a more natural environment than that demanded by other probes.

REFERENCES

1) For a review see D. POHL, Advances in Optical and Electron Microscopy, edited by Dr. C.J.R. Sheppard and T. Mulvey, to be published
2) H.K. WICKRAMASINGHE, Scientific American, p 98, October 1989
3) G. BINNIG, H. ROHRER, C. GERBER, and E. WEIBEL, Phys. Rev. Lett, 49, 57, 1982
4) R.C. REDDICK, R.J. WARMACK and T.L. FERRELL, Phys. Rev. B39, 767, 1989
5) L. SALOMON, F. DE FORNEL and J.P. GOUDONNET, proposed to Phys. Rev. B, Sept. 1990
6) J.P. GOUDONNET, F. DE FORNEL, L. SALOMON (unpublished)
7) P.J. MOYER, C.L. JAHNCKE, M.A. PIELER, R.C. REDDICK and R.J. WARMACK, Phys. Lett. A, 145, 343, 1990
8) DIN PINGTSAI, H.E. JACKSON, R.C. REDDICK, S.H. SHARP and R.J. WARMACK, Appl. Phys. Lett., 56, 1515, 1990

Combination-matching problems in the layout design of mini-laser rangefinder

Wang Erqi Song Dehui

(XSEZ Institute of Technology, Xiamen University, China)

ABSTRACT

Beginning with an analysis of the practical laser rangefinder system, this paper deals with (1) the restrictive relations between each main structural parameter of the optical system, (2)the main problem in the layout design, and (3) how to obtain the optimal design scheme of the total performance of the instrument subject to given technology by partially reducing the refractive index of the optical splitting prism material.

1. INTRODUCTION

In designing any instrument it is generally necessary to transfer functional performance indices into designing technical indices which are in turn used to determine each major structural parameter of the instrument. The most important functional performance index of the mini-laser rangefinder is its tiny volumne, light weight, convenience for field use and stability. When the parameters of such elements as mini-laser device, receiver, electric source, range-display have been determined, the key to obtaining the above-named performance index lies in the optimal combination-matching of the layout design of the instrument.

In principle, apart from an observing-aiming lens, laser rangefinder requires that at least a pair of transmitting and receiving optical devices be attached to it, functioning as antenna. Because the light-wavelengths for human eye aiming and laser operation are somewhat different, the problem concerning the combination-matching of scheme selection, structural parameters, element characteristics and material properties in the layout design are comparatively complicated. This problem is of general significance in the layout design of the instrument. The present article chiefly attempts to discuss these problems.

2. ANALYSIS OF THE MAIN POINTS IN THE LAYOUT DESIGN OF MINI-LASER RANGEFINDER

Both the observing-aiming system of the laser rangefinder and the receiving system of the backward wave purport to receive the information of the object being aimed. The demands for the parameters of the two optical system (the aperture of the objective lens for example) are, it follows, approximately similar. Normally, in the layout design, they are often combined into co-objective lens and co-image erecting systems so as to reduce the number of optical elements and the weight of the instrument. The following paragraphs discuss several systems occurred in the development of mini-laser rangefinder.

2. 1. Adopt the image erecting system of FBJ—0° complex prism (Pechan prism with roof angle) and combine the aiming system and the receiving (transmitting) system into one whole, as is shown in Fig. 1.

The main characteristics of this structure is that the ray path of the image errcting prism is relatively long and can be used to reduce the tube length. Moreover, the axises of the incoming and the outgoing rays become collinear.

The fact that the axis of transmitting, receiving and aiming systems rely on several reflectors and prisms to deflect rays gives rise to a number of unstable factors. If in operation the prisms and reflectors are slightly shifted, the parallelism of the axis will be disturbed so that the object aimed and the object measured become different, thus resulting in error measurement. Meanwhile, due to the complexity of the component links and the large size of Pechan prism, the weight of the instrument will increase, therefore violating the Principle of Mininmal Component Link.

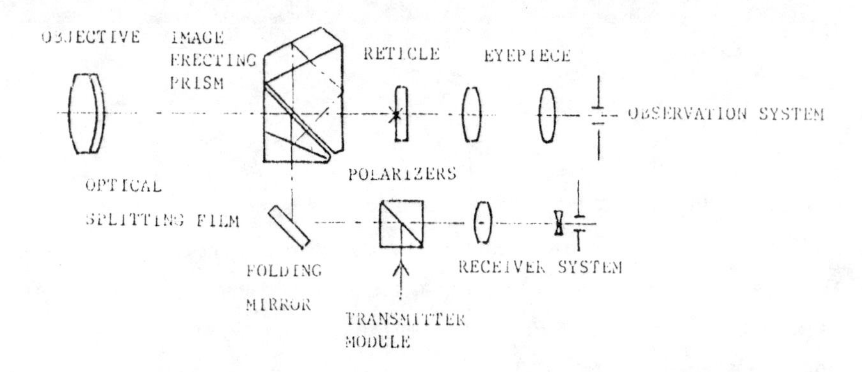

Fig. 1.

2. 2. The chart of the three axises that constitute a system on their own is diagrammed in Fig. 2, the main characteristics of which is that the observing-aiming system and the receiving system can respectively correct aberration and obtain satisfactory image quality. Once the non-spherical lens is adopted, large relative aperture can be obtained, thus greatly reducing the focal length of the objective lens and the weight of the instrument. Obviously, this system does not conform to the Principle of Minimal Component Link and will increase the volume and weight of the instrument, engendering unstable factors in the optical axis parallelism of the aiming, receiving and transmitting systems.

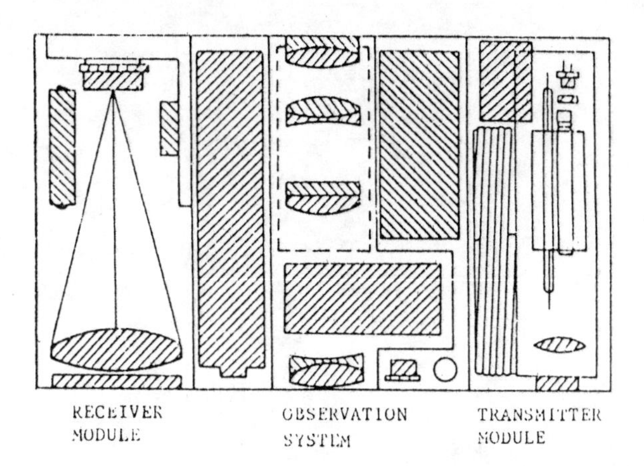

Fig. 2.

2. 3. Adopt the image erecting system with two D Ⅱ —180° vertical prisms combined (e. g. Porro model Ⅰ), making the aiming and the receiving systems into one whole as is shown in Fig. 3. The main characteristics of this structure is that shorter lens tube can be obtained by extending the distance of the two vertical prisms, under the condition that the focal length of the objective lens is longer and the image quality is better. Meanwhile the transverse size will also increase and the volume of the instrument cannot be reduced. Since the prism of the image erecting system is separative, if, in operation, of the position of the prism is slightly changed, the image quality will be affected and the parallelism of the optical axis disturbed. This structure will inevitably lead to the bulky volume of the instrument.

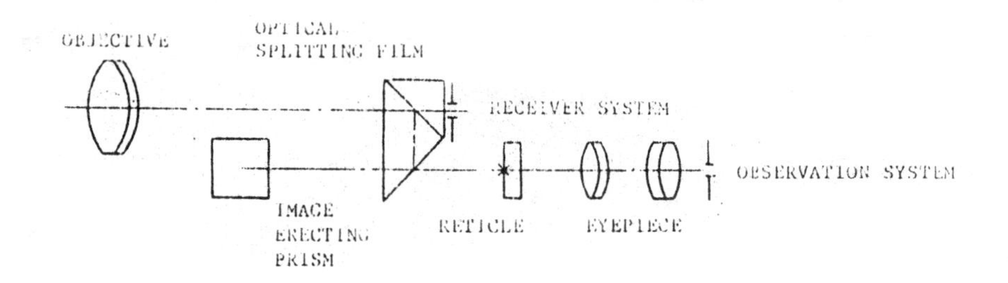

Fig. 3.

2. 4. Adopt the image erecting system with FP—O° complex prism and attach it to the focal plane of the objective lens of the telescope, making the receiving system and the aiming system merge into one, as is shown in Fig. 4. The main characteristics of this structure is that the prism is near the focus and the size is smaller. Having no additional deflect components, the system conforms to the Principle of Minimal Component Link. Because the image erecting prisms glue togother, the structure is compact and its function stable.

The optical axis of the receiving aperture and the cross centre of the aiming axis happen to be on the opposite sides of the reflector of the optical splitting prism. As a result, when the prism is unfolded according to the reflector plane the centre of the cross aiming line and the centre of the diaphragm aperture merge into a single point, creating a non-disadjustment optical system with stable functions.

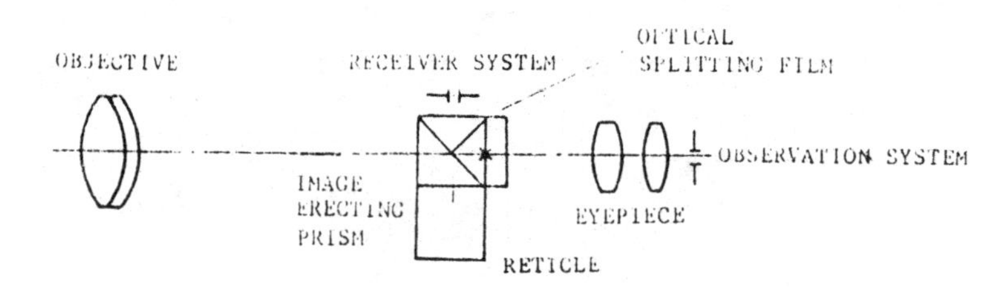

Fig. 4.

To sum up, the more reasonable method to stablize the performance function and to reduce the volume and weight of the instrument is to use FP—O° complox prisms as the image erecting system which combines the aiming with rangefinding in one construction.

3. COMBINATION-MATCHING OF PARAMETERS IN OPTICAL SYSTEM

3. 1. Main difficulty with the optical layout design

When each technical index and the structural model have been determined, the only way to reduce the volume and weight of the instrument is to shorten the focal length of the objective lens in the tele-system because of the restrictions the measuring distance of the object imposes on the aperture of the objective lens of the aiming and receiving systems. When the focal length of the objective lens is reduced, the following problems crop up: To maintain the observation field angle of the visible light, the cone angle of the incident light beam will increase with the reduction of the focal length of the objective lens. When this light beam passes through the image erecting prism the angle of incident beam facing the 45° interface of the prism will also reduce. If it is smaller than the critical angle, the internal total reflection will not occur and will result in the loss of light energy. The only way to reduce the focal length of the objective lens is, it follows, to raise the refraction index of the prism material so as to ensure the internal total reflection of the entire light beam, thus preventing the loss of the light energy received.

However the increase of the refraction index of prism material will undoubtedly lead to the intense polarization of the visible light beam of the reflection at the optical splitting films, which will in turn greatly reduce the light energy to be transmitted and seriously affect the optical splitting so that only optical splitting effect of low reflection ratio can be secured. For instance, experiments show that if the optical splitting prism made of Bak7 glass of high refraction index (n=1.5688) is used, only 60% of transmission rate can be obtained when laser beam with a wavelength of 1.06μm passes through the optical splitting film, and only 65% of reflection ratio can be obtained when the visible light beam is reflected at the optical splitting film. If the refraction index of the optical splitting prism is to be further raised the optical splitting effect will become worse and even spoil the whole operation.

Obviously, to ensure optical splitting efficiency, it is therefore necessary to reduce rather than increase the refraction index of the prism material. This means that an inharmonious contradiction arises between raising the refraction index of the prism material in order to reduce the focal length of the objective lens and reducing the refraction index of the material in order to minimize the polarization loss(i.e. to raise optical splitting effect). This is where the difficulty lies in the layout design.

3.2. Traditional solution to the problems in layout design.

In optical design the optical splitting effect is often sacrificed to the advantage of reducing the length of the tube within limits by approximately raising the refraction index of the prism material.

ZK3 glass (n=1,5890) is used accordingly. Its optical splitting effect is undoubtedly lower than that of Bak7 glass. The optical parameters of the layout design are shown in Chart. 1., from which it can be inferred that the instrument designed whereby will not only fail to reduce the weight of the instrument to the predetermined demands of performance technical index, but will also effect a decrease in the observing-aiming performance and the measuring performance of the instrument, leading to the failure of the design.

3.3. Solve the difficulty in the layout design once and for all by partially reducing the refraction index of the optical splitting prism material.

Analyses and experiments show that the material refraction index of the prism will need to increase with the decrease of the focal length, only when the incident light beam uses the interface of the prism to create internal total reflection. But, on the optical splitting film of the optical splitting cube prism, the incident beam achieves optical splitting effect by using optical splitting film to reflect and transmit one light beam with two different wavelengths. Therefore, the refraction index of the material of this small vertical prism will not need to increase with the decrease of the focal length of the objective lens. Conversely, it can select appropriate glass of low refraction index according to the needs for eleminating the reflectory polarization loss of the visible light beams. This is what is proposed here as the method of "partially reducing the refraction index of the material of the optical splitting prism".

In the optical system design the glass of lower refraction index, e.g. K3 glass whose n=1.5046 can serve as optical splitting cube prism. Experiments conducted under the condition that the coating technique (e.g. these identical with the coating technique of the optical splitting film of the prism made of Bak3 glass material) is maintained show that the transmissivity of the laser beam with a wavelength of 1.06μm can be raised to 85% while the integral reflection rate of the visible light beam can be raised to up to 80%. This is a manifestation that the combination-matching among the short focal length of the objective lens, the internal total reflection prism with high refraction index and the optical splitting cube prism (which can reduce the refraction index of the material) realizes the optimal combination-matching of the instrument structural parameters. The results of the design are also shown in Chart 1.

Comparism of two design schemes and their experiment rusults Chart 1.

Item	by traditional design method	by combination-matching method
Magnification	7x	7x
Field of view	7°	5°
Objective aperture	Φ45	Φ45
Objective focal length	200	160
Prism size	26. 5×26. 5	18×18
Eye piece aperture	Φ36	Φ23
Eye piece type	5pieces	4 pieces, symmetry type
Total length of telescope	240	200
Material of the optical splitting prism	ZK3 (n=1. 5890)	K3 (n=1. 5046)
Optical splitting effect: transmission 1. 06μm reflection visible light	less60% less65%	85% 80%
Weight	3. 5kg	2kg

4. CONCLUSION

The principle of the optimal combination-matching of the structural parameters has been successfully applied to the layout design of Chinese miniature laser rangefinders and has successfully made the instrument so designed meet the performance demands of tiny volume, light weight, convenience for field use and stability.

With the development of laser technology, 0. 5kg mini-laser rangefinders have come into being. This reality poses higher demands for the combination-matching of the instrument layout design.

5. REFERENCES

1. Tang Jiuhua, "On Problems Towards The Promotion of Instrument Layout Design", ACTA OPTICA SINCA, Vol. 1, No. 1, Shanghai, Jan. 1981.

2. Wang Erqi, Song Dehui, ACCURACY ANALYSIS OF OPTICAL INSTRUMENT, Survey & Drawing Publishing House, Beijing, Oct. 1988.

3. Schlecht, G Richard, Jeffrey L. Paul , ADVANCES IN MINIATURE FOR RANGING APPLICATIONS, SPIE, Vol. 247.

Design and testing of a cube-corner array for laser ranging

W E James

James Optical Pty Ltd, 19 Rosslyn Street
East Hawthorn, Vic., Australia 3123

W H Steel

School of Mathematics, Physics, Computing, and Electronics
Macquarie University, NSW, Australia 2109

N O J L Evans

British Aerospace Australia
PO Box 180, Salisbury, SA, Australia 5108

ABSTRACT

A *Laser Retroreflector Unit*, consisting of an array of cube corners, has been designed and made for the satellite *Aussat B*. The design involved the study of cube corners with small angle offsets, the diffraction pattern they produce, the effects of surface coatings, and their transmittance at high angles of incidence. Interferometric tests were developed to measure during manufacture the angle offsets of the cube from which the retroreflectors were cut, and radiometric tests confirmed the final performance.

1. INTRODUCTION

The satellite *Aussat B* is to carry a cube-corner array as part of a laser ranging system, to be used to synchronize atomic clocks in Australia and New Zealand. The satellite will carry mainly microwave equipment and the retroreflector must not affect this.

Because of the rotation of the earth during the passage of light to the satellite and back, the satellite will see the receiver at an angle of 18 μrad (3.7") from the source. Since the distribution of light from a retroreflector is spread over all azimuths, the ideal system is one that returns the incident beam as a cone with 18 μrad semi angle.

From a comparison with other retroreflectors from which signals have been received, it was specified that the retroreflector unit should have a *specific intensity*

$$I_s = 10^6 \text{ mm}^2 \text{ sr}^{-1}, \qquad (1)$$

where *specific intensity* is a convenient quantity for describing a retroreflector, being the reflected radiant intensity divided by the irradiance at its surface.

2. ANGLE OFFSETS

When the dihedral angles between the faces of a cube corner depart from 90°, an incident beam leaves as six separate beams, the direction of each depending on the order in which it meets the reflecting surfaces and on the angle offsets. The relation between the beam angles to the incident beam and these offsets has been given by Yoder[1] and this shows that the six beams leave at angles of 3.7" to the incident beam if the angle offsets are ±0.8" and all have the same sign.

For testing it is desirable to be able to find each offset angle from the six beam directions. Yoder's treatment is not sufficient for this without some extension. With unit vectors **i**, **j**, and **k** in the directions of the edges of a cube corner with no angle errors, small angle offsets α, β, and γ, refractive index n, and an incident ray

$$\mathbf{r} = -(\mathbf{i} + \mathbf{j} + \mathbf{k})/\sqrt{3}, \tag{2}$$

the rays reflected from the first three sectors (Fig. 1a) are

$$\begin{aligned}
\mathbf{r}'_1 &= n[(1 + 2\beta + 2\gamma)\mathbf{i} + (1 - 2\gamma + 2\alpha)\mathbf{j} + (1 - 2\alpha - 2\beta)\mathbf{k}]/\sqrt{3},\\
\mathbf{r}'_2 &= n[(1 + 2\beta - 2\gamma)\mathbf{i} + (1 + 2\gamma + 2\alpha)\mathbf{j} + (1 - 2\alpha - 2\beta)\mathbf{k}]/\sqrt{3},\\
\mathbf{r}'_3 &= n[(1 - 2\beta - 2\gamma)\mathbf{i} + (1 + 2\gamma + 2\alpha)\mathbf{j} + (1 - 2\alpha + 2\beta)\mathbf{k}]/\sqrt{3}.
\end{aligned} \tag{3}$$

Those from the sectors opposite these have opposite deviations.

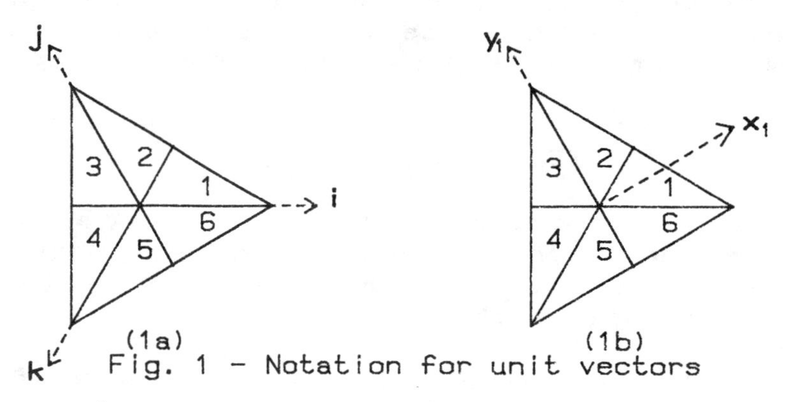

Fig. 1 – Notation for unit vectors

It is more convenient to refer these ray directions to unit vectors in the exit face of the cube corner (Fig. 1b). Different vectors are used for each pair j of sectors: a unit vector $\mathbf{x}_j$ bisects the sector angle, and $\mathbf{y}_j$ is normal to it. The differences between the directions of rays from opposite sectors can be expressed as

$$\mathbf{r}'_{3+j} - \mathbf{r}'_j = \rho_j \mathbf{x}_j + \sigma_j \mathbf{y}_j,$$

where

$$\begin{aligned}
\rho_1 &= 2n\sqrt{2}(2\alpha + \beta + \gamma)/\sqrt{3}, & \sigma_1 &= 2n\sqrt{2}(\gamma - \beta),\\
\rho_2 &= 2n\sqrt{2}(\alpha + 2\beta + \gamma)/\sqrt{3}, & \sigma_2 &= 2n\sqrt{2}(\alpha - \gamma),\\
\rho_3 &= 2n\sqrt{2}(\alpha + \beta + 2\gamma)/\sqrt{3}, & \sigma_3 &= 2n\sqrt{2}(\beta - \alpha).
\end{aligned} \tag{4}$$

The six components ρ_j and σ_j give six equations for the three angle offsets.

A least mean square solution is

$$\alpha = (\sqrt{3}\rho_1 - \rho_2 + \rho_3)/10 + A,$$
$$\beta = (\rho_1 + \sqrt{3}\rho_2 - \rho_3)/10 + A,$$
$$\gamma = (-\rho_1 + \rho_2 + \sqrt{3}\rho_3)/10 + A,$$

where

$$A = \sqrt{3}(\rho_1 + \rho_2 + \rho_3)/20. \quad (5)$$

3. DIFFRACTION THEORY

The complex amplitude in the far field of a circular aperture of radius a is given by

$$A(r,\theta) = \frac{a}{\lambda\sqrt{\pi}} \int_0^1 \rho d\rho \int_0^{2\pi} \exp\{-i\rho[r\cos(\theta-\gamma)]\}d\gamma, \quad (6)$$

where the expression is normalized to give unit intensity when integrated over all angles. The quantity r is related to the deviation angle R by

$$r = 2\pi aR/\lambda, \quad (7)$$

θ is the azimuth of the reflected ray, and ρ and γ are polar co-ordinates at the aperture; $\rho = 1$ at its edge. For a perfect reflector,

$$A(r,\theta) = \frac{a\sqrt{\pi}}{\lambda} 2 \frac{J_1(r)}{r}, \quad (8)$$

and the central intensity is $I_0 = |A(0)|^2 = \pi a^2/\lambda^2$. (9)

For an array of N cube corners the specific intensity is

$$I_s = \pi^2 N a^4/\lambda^2. \quad (10)$$

The *Aussat B* unit consists of 14 cube corners, each 38.1 mm diameter and equation (10) gives a specific intensity on axis of a perfect array with no angle offsets as 6.44×10^7 $m^2 sr^{-1}$ for $\lambda = 532$ nm.

Chang *et al.*[2] have shown the need for calculating the complex amplitudes from each sector separately. Their treatment for a cube corner with no angle offsets can be applied readily to the case of offsets. The complex amplitude from one sector, relative to that (8) at the centre of an Airy disk, is given by a modification of (6) as

$$a_i(r,\theta) = \frac{1}{\pi} \int_0^1 \rho d\rho \int_{-\pi/6}^{+\pi/6} \exp i\{-\rho[r\cos(\theta-\gamma) - g\cos\gamma]\}d\gamma. \quad (11)$$

This takes the same form as the expression used by Chang *et al.* if the variables are changed to

$$\sigma = (r^2 - 2rg\cos\gamma + g^2)^{1/2}, \quad \tan\Psi = r\sin\theta/(r\cos\theta - g). \quad (12)$$

Their treatment then gives

$$a_i(r,\theta) = J_1(\sigma)/3\sigma + (2/\pi)\sum_{k=1}^{\infty}(-i)^k \sin(k\pi/6) \cos k\psi\, F_k(\sigma), \qquad (13)$$

where

$$F_k = \frac{1}{kx^2}\int_0^x x\, J_k(x)\,dx,$$

$$= (1/k) \sum_{m=0}^{\infty}(-1)^m(x/2)^{k+2m} [(k+m)!\,m!\,(k+m+1)]^{-1}. \qquad (14)$$

If polarizations are all the same, the amplitude from the whole cube corner is

$$A(r,\theta) = \sum_{j=0}^{5} a_i(r, \theta + j\pi/3). \qquad (15)$$

Since the array is illuminated by a laser, the amplitudes from all the cube corners add. But their relative phases are unknown. Trial summations with random phases give typical speckle patterns. To estimate the average return signal, a summation of the intensities from each cube corner is taken. For the array of 14 cube corners of 38 mm diameter, this average at 0.18 μrad (3.7") off axis is

$$5\mathrm{x}10^6\ \mathrm{m}^2\ \mathrm{sr}^{-1}. \qquad (16)$$

4. POLARIZATION

Equation (16) applies when all six beams have the same polarization. But Chang *et al*. have pointed that the beam polarizations from an uncoated cube corner of fused silica are very different and cause a wider spread of the reflected light than the Airy disk. They show that coating the reflecting surfaces with either silver or aluminium will reduce this. To meet the microwave requirements, the cube corners have a conducting coating on their front faces and it was investigated whether this could be used also on the reflecting faces to give a similar effect on the polarization.

A combination of the expressions for total reflection from a boundary between a medium of refractive index n and air[4],

$$r_{23p} = \frac{a^2\cos\theta_2 - i(\sin^2\theta_2 - a^2)^{1/2}}{a^2\cos\theta_2 + i(\sin^2\theta_2 - a^2)^{1/2}}, \qquad (17)$$

$$r_{23s} = \frac{\cos\theta_2 - i(\sin^2\theta_2 - a^2)^{1/2}}{\cos\theta_2 + i(\sin^2\theta_2 - a^2)^{1/2}}, \qquad (18)$$

where $a = 1/n$ and θ_2 is the angle of incidence within the coating, and that for a two-layer system[4]

$$r = \frac{r_{12} + r_{23}e^{-2i\beta}}{1 + r_{12}r_{23}e^{-2i\beta}} \tag{19}$$

is used, where $\beta = (2\pi nd/\lambda)\cos\theta$. This gives a maximum phase difference between the principal polarizations for an indium-tin oxide coating (ITO, $n = 2$) on silica at an angle of incidence of 54.74° and λ = 532 nm when the coating has a thickness d of 112 nm. Table I compares this coating with metals. The term $|\xi|$ is the measure used by Chang *et al.* to indicate the amount of light in the same state of polarization, R is the reflectance of the system (three reflections), and the phase difference Ψ should be 180° for a perfect reflector. A figure of merit is provided by the product $R|\xi|^2$.

Table I

Coating	$\|\xi\|$	Ψ	R	$R\|\xi\|^2$
none	0.52	42°	1.00	0.27
ITO	0.95	109°	1.00	0.90
Ag	0.95	121°	0.94	0.84
Al	0.80	137°	0.66	0.43

The promising coatings are ITO and silver. A coating of ITO was used; at the optimum thickness this will the reduce the theoretical specific intensity of (16) to $4.5\text{x}10^6$ mm^2 sr^{-1}, to which is added small contributions from the unmatched polarizations, as given by Chang *et al*[2].

5. TRANSMITTANCE

When a ray is incident at a reflecting face of a cube corner at an angle γ greater than 43.2°, it is transmitted and this transmitted radiation can heat the mounting. This angle δ depends on both the angle of incidence at the front face α and also the azimuth γ of the ray; it is given by

$$\cos\delta = (2/3)^{1/2}\sin\theta\cos\gamma + (1/3)^{1/2}\cos\theta, \tag{20}$$

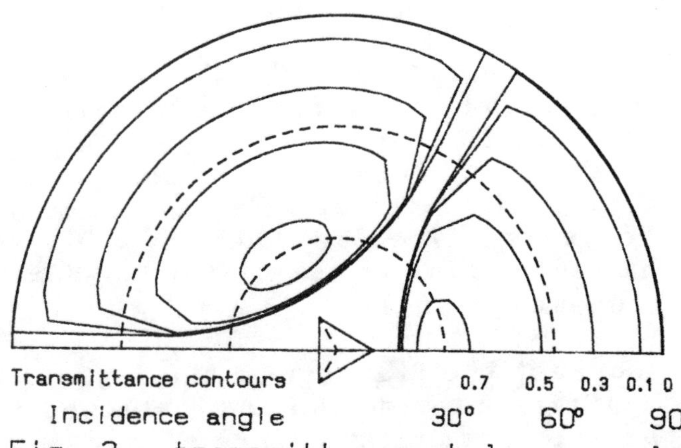

Fig. 2 - transmittance at large angles

where θ is the angle of refraction at the front surface, $n\sin\theta = \sin\alpha$. Radiation is transmitted when $\sin\delta \geq 1/n$ with a transmittance that is the product of that through the front surface, which has an ITO coating and an anti-reflecting coating, the transmittance through the ITO coating on the rear surface, and an obliquity factor, $\cos\alpha$. Transmittance contours are given in Fig. 2 for azimuths from 0° to 180°.

6. AN INTERFEROMETER TO MEASURE RIGHT ANGLES

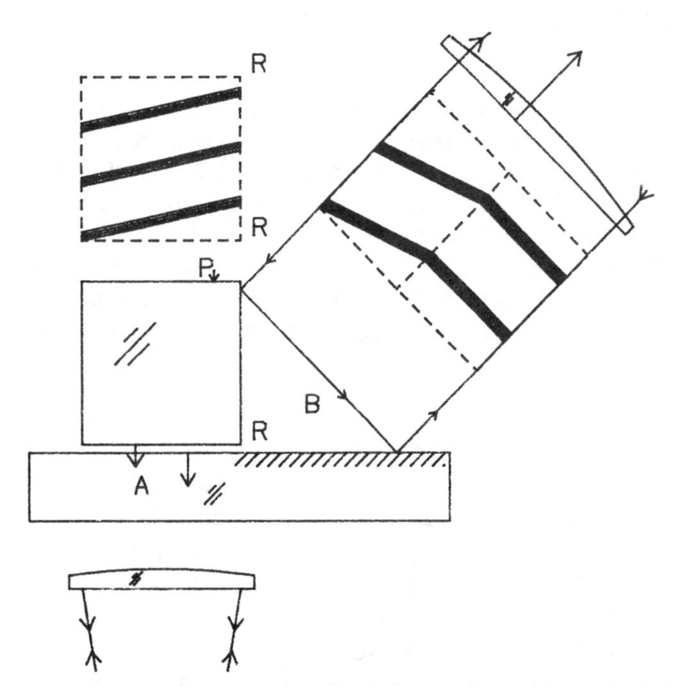

Fig. 3 - angle test, showing fringes seen at A and B

The cube corners were cut from 50 mm silica cubes, polished to the correct angle and an interferometer was developed to measure the angles. In Fig. 3 is shown the plan view of the master optical flat (an angle standard of 180°) and a 50 mm cube. The interference fringes at A were viewed by a Fizeau interferometer with a He-Ne laser at 543 nm as source.

The angle at B was viewed using a second Fizeau interferometer. For this the collimating lens was a single plano-convex lens with the flat face towards B acting as a reference flat. Since the optical path difference between this flat and the angle B was about 300 mm, the light source for this interferometer was a single-mode stabilized He-Ne laser at 633 nm. One half of the master flat was aluminized to obtain maximum fringe contrast. Both it and the reference surface of the lens were flat to $\lambda/50$.

6.1. Interpretation of fringes

Typical fringe patterns are shown in Fig. 3. The tilt of the fringes A give the air-gap angle between the cube and the master flat, while the deviation of the fringes in view B is a measure of the supplement of the cube angle. When the fringes A are normal to the edge R-R and there is no deviation of the fringes B, the angle of the cube is exactly 90°. Then a change to the air-gap angle will give a tilt to the fringes A and the same deviation to B.

The deviation of the cube angle from 90° can be measured by adjusting the tilt of the cube relative the master flat until either the fringes A show no tilt or the fringes B no deviation. The other fringes then measure the dihedral angle offset. The number of fringes in each set is chosen to give a suitable sensitivity. The sign of the angle offset is found by applying slight pressure on the cube against the master flat at some point P opposite the edge R-R. This reduces the supplementary angle at B.

During polishing the cube faces to correct the flatness or the angle, the temperature of the cube is raised and sufficient time had to be allowed for it to reach the same temperature as the interferometer before measurement, particularly for the final critical measurement.

6.2. Image acquisition

The two fringe images were viewed by CCD cameras and stored sequentially through a frame grabber with a time interval of 0.5 s. They were then adjusted in intensity and contrast, cropped when necessary, and printed. Measurements on the printed interferograms gave the angle offset to ±0.5 μrad (0.1").

The finished cube corners were tested on a Wyco phase-stepping interferometer at the CSIRO National Measurement Laboratory and the results showed good agreement with those obtained with our system. This interferometer also showed the quality of the wave fronts leaving each sector.

7. RADIOMETRIC TESTS

To measure the specific intensity of the array, the equipment shown in Fig. 4 was set up. It has a collimating lens of 110 mm diameter and 2.7 m focal length; with this diameter measurements can be made over three cube corners of the array. The objective is a separated doublet with both components of BK 7 glass, corrected for spherical aberration but not for colour. A magnifying lens increases the equivalent focal length to 103 m, so that an angle of 1" gave an image of 0.5 mm. At the low numerical aperture, a sphere is sufficiently well corrected and avoids squaring-on problems.

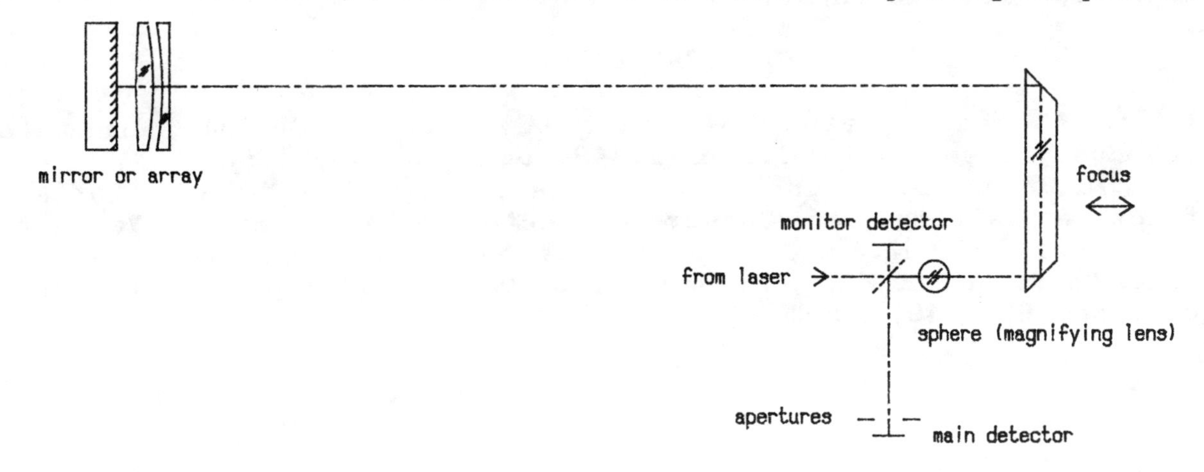

Fig. 4 - layout of test bench

The array is illuminated by a laser through a 0.5 mm pinhole and the return radiation goes via a beam splitter to a series of circular apertures that allows it to be measured over solid angles of increasing angular radii: 1", 3.4", 3.7", and 4". These give, by differences, the intensity at 3.7" and its variation around this.

The irradiance at the array is measured by replacing the array by an aluminized plane mirror, centring and focusing the return image, measuring the total return radiation over an aperture of 10" radius, and correcting for the amount of the Airy disk that is outside 10". The mirror is covered by an aperture with three holes the size and position of the cube corners. This corrects for the drop off of the radiation laser towards the lens edge.

As a test of the system, the specific intensities from the mirror were measured for circular apertures of different sizes and compared with the theoretical values for an Airy disk.

Variations of laser power and amplifier gain are monitored by a second detector in the fourth arm and ratios recorded. Most measurements have been made with chopped radiation from a green He-Ne laser at 543 nm. They agree closely with check readings at 532 nm, the wavelength at which the array will be used.

The measurements showed that the specific intensity meets the specified value but is considerably less that predicted by theory. But, because the measurements have to made over such small angles, the equipment is very sensitive to small disturbances. Radiometric measurements are best seem as confirmation only; the interferometer test is a more reliable predictor of quality.

8. ACKNOWLEDGEMENTS

This work was performed for British Aerospace Australia Ltd., who are contractors to Hughes Aircraft Corporationtion, the manufacturers of the satellite, *Aussat B*. The array will be used by the Orroral Geodetic Observatory and the authors wish to thank J.McK. Luck there for his many contributions to the theory.

9. REFERENCES

1. P. R. Yoder, "Study of Light Deviation in Triple Mirror and Tetrahedral Prisms," *J.Opt.Soc.Am.* **48**, pp. 496-499, 1958.
2. R.F. Chang, D.G. Currie, C.O. Alley, and M.E. Pittman, "Far-Field Diffraction Pattern for Corner Reflectors with Complex Reflection Coefficients," *J.Opt.Soc.Am.* **61**,pp. 431-438, 1971.
3. M. Born, and E. Wolf, *Principles of Optics*. London: Pergamon Press, London, pp. 65 & 49, 1959.

The study on the mode and far-field pattern of diode laser phased arrays

Zhang Yueqing,Wu Shengli,Zhu Lian,Zhang Xitian,Piao Youzhi,Li Diaen.

Changchun Institute of Physics,Chinese Academy of Sciences,
Changchun,130021,Jilin,P.R.China.

ABSTRACT

Diode laser phased array of single-mode,single-lobe and diffraction limited far-field pattern is pursued for many applications.According to coupled-mode analysis three types **diode laser phased** arrays **have been** designed and fabricated.The properties of these three types arrays show that the asymmetrical linearly chirped array can fullfill the application demands.

1.INTRODUCTION

Semiconductor lasers have many advantages:small size,high efficiency ,vibration-proof etc..But the relatively smaller optical output power limits its applications in some areas such as optical ranging,optical pumping of solid-state laser,eye surgery,lasercomp,optical disk for recording.One methode to develop high power semiconductor laser is to fabricate diode laser array.Phase-locked array is a multi-channels waveguide device.In general it supports many lateral-modes(supermodes) and this results in broad far-field pattern as well as broad spectral linewidth.The way to discriminate among the supermodes and enforce fundamental supermode exciting is pursued.If fundamental supermode is operating,the high quality optical beam of single-mode and single-lobe with diffraction-limited far-field pattern will be obtained. The purpose of this paper is to design the structure of array by coupled-mode analysis and experimental methode.

2.THEORY

Usually the diode laser phased arrays are analysed using coupled-mode theory.[1]
If the array consists of N waveguides with equal width d and equal spacing s.That is an uniform laser array.Assuming the position of element n is located at s_n.Then the aperture size of the array is $s_1 - s_n = N\times s$.The elements are assumed to be identical,so the field satisfy the condition $\psi^1(x,s_1,z) = \psi^2(x,s_2,z) = \ldots = \psi^N(x,s_N,z)$.
The optical field of single element satisies the wave equation:

$$\nabla^2\psi^m + k_o k^m(x,y)\psi^m = o, \tag{1}$$

k_o is the free space propagation constant and $k^m(x,y)$ the dielectric function that describes the individual waveguide.The wave function is written as

$$\psi^m(x,y,z) = U^m(x,y)\ V^m(y)\exp(-\gamma_m z), \qquad (2)$$

where $U^m(x,y)$ describes the transvere (perpendicular to the junction) profile while V^m governs the lateral mode profile. γ_m is the propagation constant of mth element.
Assuming the fields are polarized along y,we have for the individual laser (mth element)

$$E_{ym} = U^m(x,y)V^m(y)\exp(-\gamma_m z) = e_{ym}(-\gamma_m z). \qquad (3)$$

According to the coupled-mode analysis the total field is a linear combination of fields in the individual lasers.

$$\vec{E} = \sum_m A^m(z)E_m. \qquad (4)$$

$$\vec{H} = \sum_m A^m(z)H_m. \qquad (5)$$

and $$dA^n/dz = k_o^2/2 \sum_m \frac{1}{\gamma_n} c_{nm} A^m. \qquad (6)$$

c_{nm} is the coupling coefficient between nth and mth element.
From calculation we can get

$$\gamma_p = \alpha + j\beta - (k_o/n_e)\ [c_i - jc_r]\ \cos(p\pi/N+1) \qquad (7)$$

$$(\lambda_p - \lambda_o)/\lambda_p = -c_r/n_e^2\ \cos(p\pi/N+1) \qquad (8)$$

$$G_p = G + (2k_o c_i/n_e)\ \cos(p\pi/N+1) \qquad (9)$$

G_p is the pth modal gain, $n_e = \beta/k_o$, $\gamma = \alpha + j\beta$, $G = -2\alpha$, γ^p and λ_o is the propagation constant and wavelength of single element respectively.
For Uniform array $c_{nm} = c = c_r + jC_i$.

If the array is nonuniform with varying waveguide width and varying spacing,the modal gain,wavelength and propagation constant can not get from the solution of coupled-mode analysis exactly.But the super mode gain can be calculated numerically.
The resulting relative modal gain of three types arrays are shown in Fig.1.[2]

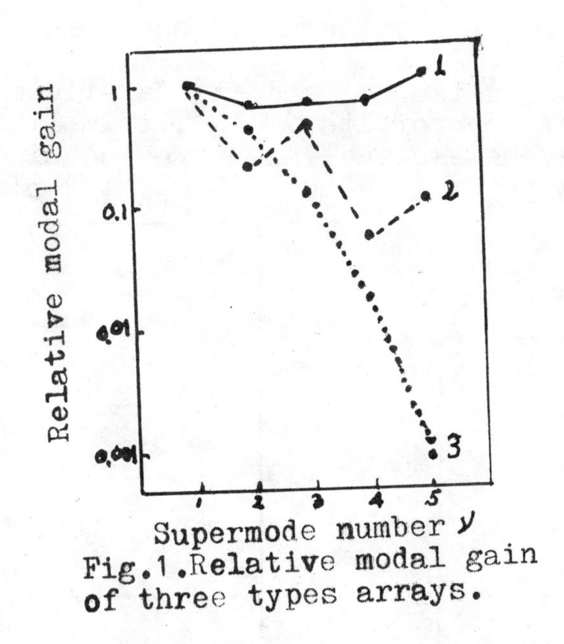

Fig.1.Relative modal gain of three types arrays.

Curve 1 represents the modal gain distribution among supermodes of uniform array.It shows that the modal gain of fundamental mode and the highest supermode are almost equal but the highest supermode gain is slightly larger,so the highest supermode will be excited at first.

Curve 2 represents the modal gain distribution of symmetrical nonuniform array with varying waveguide width and constant spacing, i,e,inverted"v" chirped array.It shows that fundamental super mode modal gain is higher than the other supermodes.But the difference of the modal gain between fundamental supermode and third supermode is not bigger enough.Sometimes the third supermode may be excited.

Curve 3 represents the modal gain distribution of linearly chirped array.It shows that the difference of modal gain between super modes is very large and the modal gain of fundamental super mode is largest.Hence the fundamental supermode will be excited mostly.Then this will operate stable single mode with single lobe diffraction-limited far-field pattern.

3.THE STRUCTURE OF ARRAY AND ITS PROPERTIES

The double heterostruture laser wafers of arrays are grown by LPE (liquid phase epitaxy).The laser array strip pattern are formed by two-steps proton bombardment.The cross section of the array is shown in Fig.2.

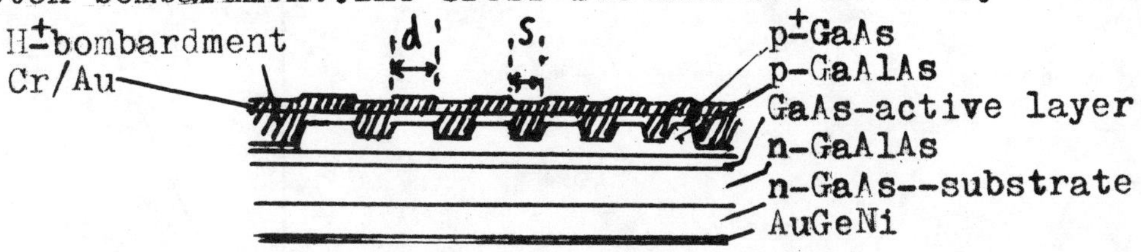

Fig.2 The cross-section of array.

The strip width of uniform array is 4 um and the spacing is 6 um. The spectrum and the far-field pattern are shown in Fig.3 and 4 respectively.

This array is excited in multimodes and its far-field pattern is double-lobes without zero center. Its $\theta_{\parallel} \simeq 10^{0}$. It implies that the highest and fundamental mode may be excited. This array consists of five elements. The optical output power is about 150 mW at $2.5I_{th}$. shown in Fig. 5.

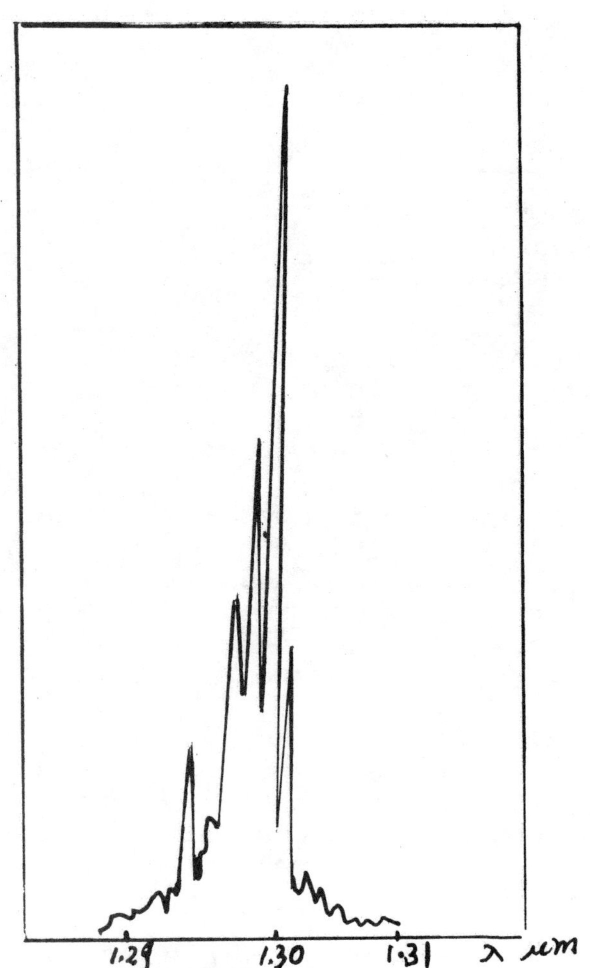

Fig.3. The spectrum of uniform array.

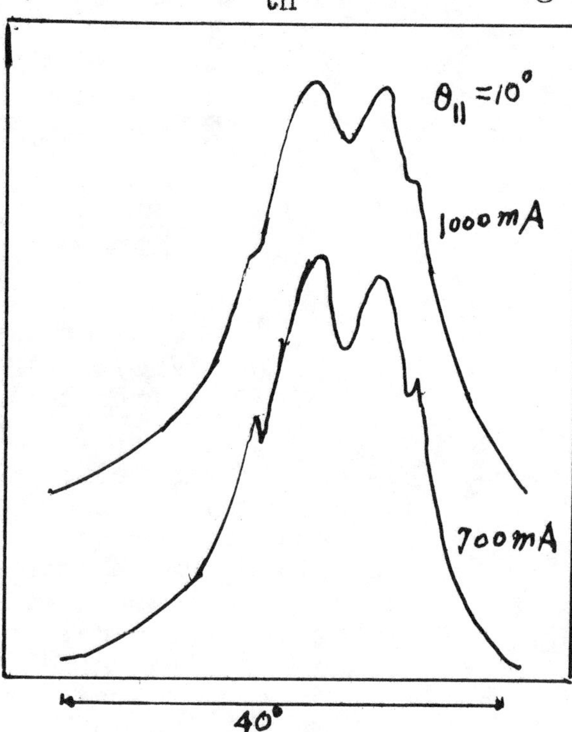

Fig.4. The far-field pattern of uniform array.

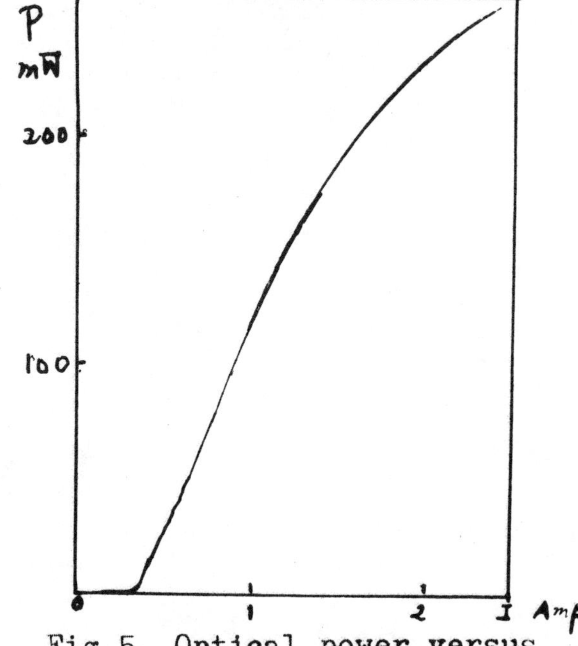

Fig.5. Optical power versus injection current.

The strip width of symmetrical "v" chirped array is 2,3,4,4,3,2 um. Its spacings are 3 um.The spectrum of symmetrical chirped array is shown in Fig.6. Under threshold cuurent it is single mode operation. It goes to double mode as soon as injection current approaches to 1.6 I_{th}.The far-field pattern is shown in Fig.7.It is single-lobe pattern with $\theta_{\parallel}=5^{o}$.

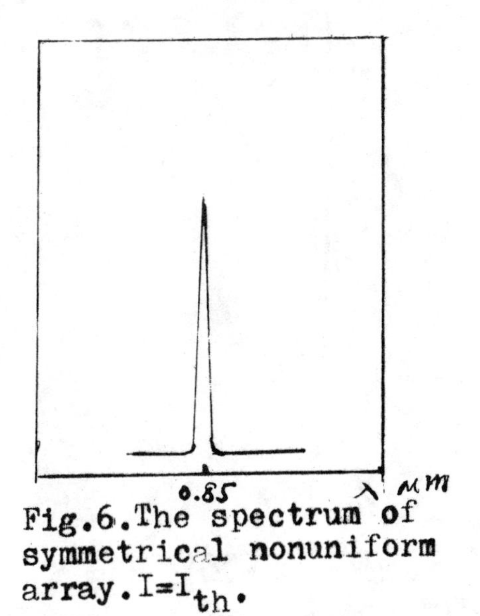

Fig.6.The spectrum of symmetrical nonuniform array. $I=I_{th}$.

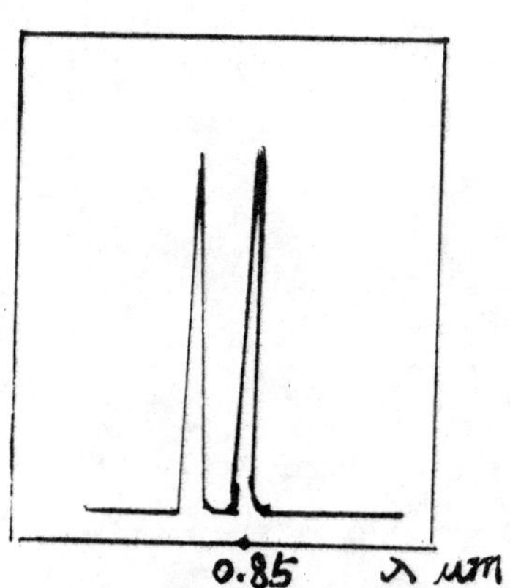

The spectrum of symmetrical nonuniform array. $I=1.6I_{th}$.

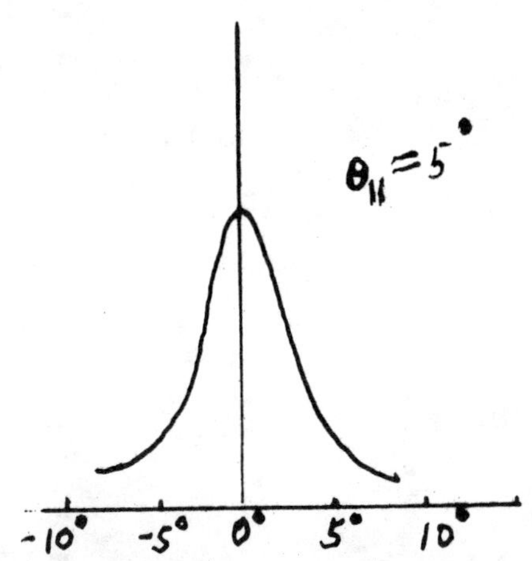

Fig.7.The far-field pattern of symmetrical nonuniform array.

The width of stripes of linearly chirped asymmetrical array is 3,4, 5,6,7,um.while its spacings equal to 5 um.The spectrum of linearly chirped asymmetrical array is shown in Fig.8.It is single mode. The far-field pattern is shown in Fig.9.This is a single-lobe pattern with diffraction-lomited $\theta_{\parallel} = 1.9^{\circ}$.Its optical output power versus injection current shown in Fig.10.The output power is about 300 mW when injection current equal to $2.8I_{th}$.

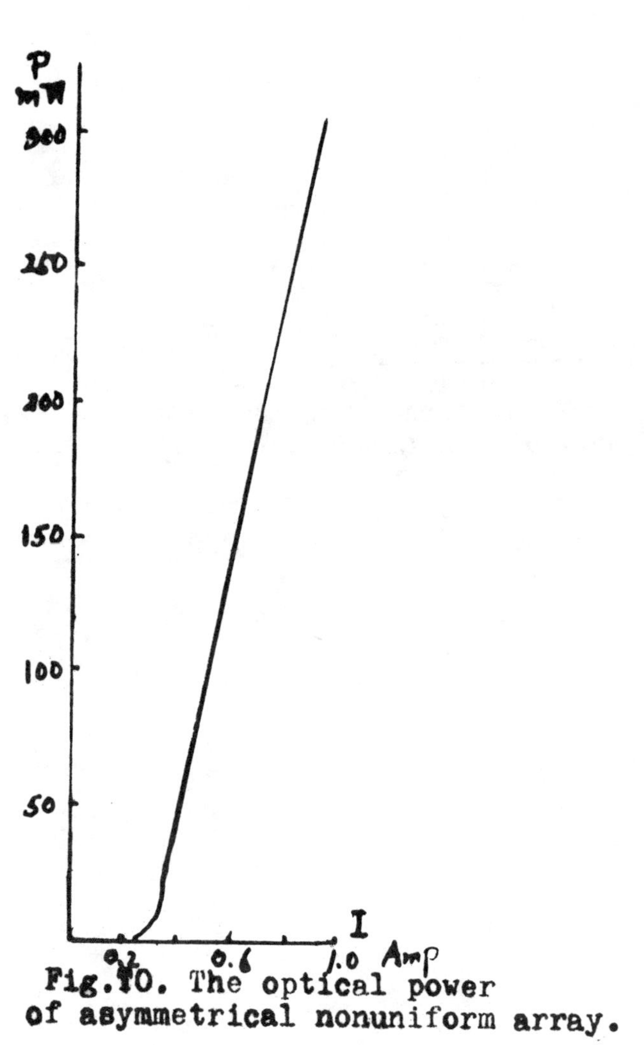

Fig.10. The optical power of asymmetrical nonuniform array.

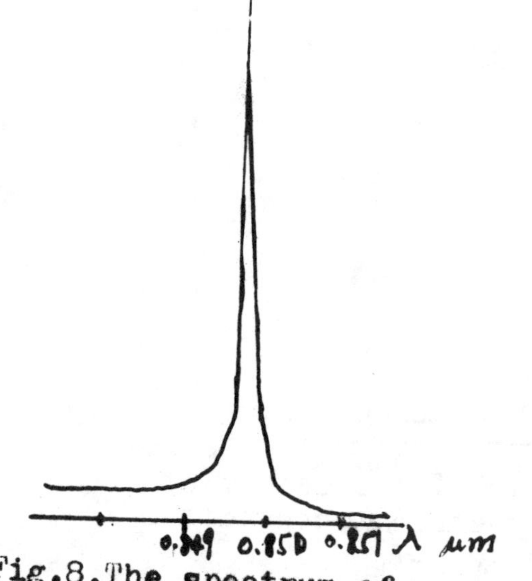

Fig.8.The spectrum of asymmetrical nonuniform array.

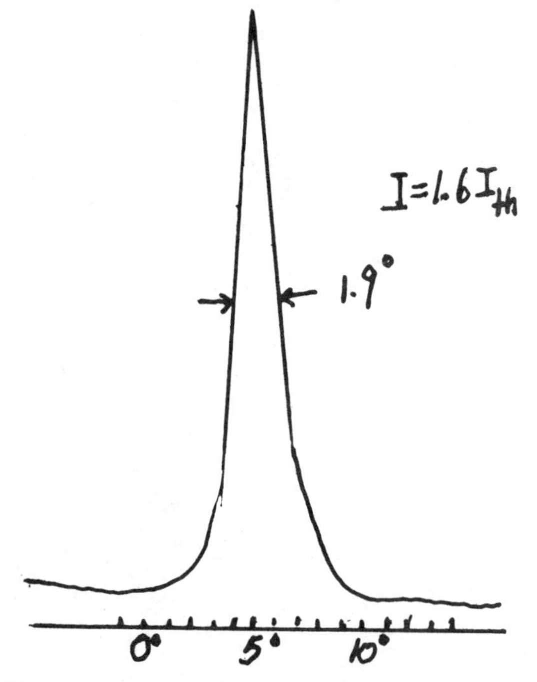

Fig.9.The far-field pattern of asymmetrical nonuniform array.

4.CONCLUSION

The experimental results and coupled-mode analysis both show that linearly chirped asymmetrical diode laser array often operates in fundamental supermode.Its optical beam is good in quality with single-lobe diffraction-limited pattern.Its weakness are small aperture size and optical output power is about hundreds mW only.

5.REFERENCES

1. Jerome K.Butler,Donald E. Ackley and Michael Ettenberg,"Coupled-mode analysis of gain and wavelength oscillation characteristics of diode laser phased arrays",IEEE J.of Quantum Electronics,Vol. QE-21,No.5,pp.458-464,May 1985.
2. E.Kapon,C.Lindsey,J.Katz,S.Margalit and A.Yariv,"Chirped arrays of diode lasers for supermode control",Appl.Phys.Lett.Vol.45,No.3,pp. 200-202,August 1984.

SESSION 4

Fabrication and Manufacture

Chair
Manfred Lorenzen
Wild Leitz Singapore Pte. Ltd. (Singapore)

NEW GLASSES FOR OPTICS AND OPTOELECTRONICS

Hans F. Morian

SCHOTT GLASWERKE, Mainz/FRG

ABSTRACT

Optical glasses, due to their high homogeneity regarding physical properties, and the ability to tailor key characteristics to meet the desired performance, find new applications in the wide fields of optics and optoelectronics.

1. In the optical glass field the general development goals are:

   1.1 - optical glasses with anomalous partial dispersion characteristics for apochromatic colour corrections

   1.2 - glasses with reduced weight, so called lightweight glasses

   1.3 - glasses with improved UV-transmittance

   1.4 - new glass types for deep UV-applications

   1.5 - optical glasses with high homogeneity in diameter up to 1555 mm

   1.6 - highly prestressed windows dia. 930 mm for bubble chambers.

   the latest results will be presented.

2. For radioactive environments and for space applications a variety of stabilized optical glasses have been developed. These glasses have been tested with electron-, proton-, neutron- and gamma radiation.

3. Special glasses exist for the use in Cerenkov and Scintillation counters for particle research.

4. Nonlinear optical effects in semiconductor doped glasses are under investigation.

1. New optical glasses for imaging optics.

In figure 1 all new types are marked.

1.1 New optical glasses with anomalous partial dispersion.

To obtain a high image quality, all the image defects will have to be corrected in an optimum way. All image defects, leaving aside the chromatic aberrations, can be corrected by combining different types of glass. To eliminate the chromatic defect, a crown glass employed as a convergent lens, for example, can be combined with a flint glass lens used as a divergent lens, by which an equal focal point is achieved for two colours. This system is known as an achromat. The lack of chromatic definition (secondary spectrum) remaining, however, still is a disturbing factor in most of the applications. The secondary spectrum will be corrected if special glasses, namely long-crown glasses (FK, PK, PSK glasses) and short-flint glasses (KzFS) are employed. If equal focal points are achieved for three or more colours, we call this an apochromat and a superachromat, respectively. The larger the focal lenght, the more severely will be noticed the residual chromatic aberration. Examples of lens systems with improved chromatic corrections are shown in figure 2.

It starts from prior art glasses, i.e. BK 7 and F 2; by replacing the crown glass with FK 51 one achieves an apochromat and by further use of PK 51A and KzFS types the colour correction of this system is further improved.

In the long-crown region a new glass, named PSK 54, with nd = 1.5860 and vd = 64,1 has been developed. Existing glass types i.e. PK 51 A and PSK 53A show improved properties like higher internal transmittance.

In the short-flint region the glass KzFS 7A was redeveloped with the result of doubling the short flint character, making this glass more interesting for apochromatic colour corrections.

1.2 New lightweight glasses

3 new glasses have been added to the family of lightweight glasses: (the percentage value shows the weight reduction compared with existing glass)

SF L 4	30 % lighter than	SF 4
SF L 57	35 % " "	SF 57
LaSF 36 A	20 % " "	LaSF 33

In comparison with the existing glass types, which contain lead oxid (in the case of F and SF types) or other heavy elements (in the case of LaSF) these elements have been replaced with lighter oxids like Titaniumoxid and others. Besides the weight reduction, I want to emphasize that all these new glass types show substantial improvements in chemical resistance, a higher Tg (transformationpoint) value and a greater hardness.

Therefore these glasses are easier to polish and coat, enabling reduced manufacturing costs.

Figure 3

1.3 Glasses with improved UV-transmittance

Improving the UV-transmittance of optical glasses is a steady task. This can be achieved by using raw materials with smaller amounts of impurities (like colouring agents), corrected melting techniques and a change of refining agents, while maintaining cost effective production.

For the wavelengths of 365 nm and 400 nm at a sample thickness of 25 mm the improvements of the internal transmittance for some of our catalogue glasses are listed; the old values have been achieved 1980. The new improved results were reached during the past years.

See table No. 4

All these glasses are now in regular production and supply.

1.4 New glass types for deep UV-applications

For this region up to 200 nm optical designers could use only fused silica and some crystals for the necessary colour correction. (Commonly-used crystals are CaF_2, MgF_2 and LiF).

Instead of crystals with their draw backs optical designers want homogeneous optical glass types.

SCHOTT has worked on new melting techniques in proprietory melting units, to produce special glasses with transmission up to 200 nm.

These new fluorophosphate glasses are named "ULTRAN" and have low refractive indices and high Abbe-values.

Figure 5 shows the transmission of ULTRAN 10, 20 and 30 in comparison with typical optical glasses.

A final data sheet has been published for ULTRAN 30; ULTRAN 10 and 20 will follow in a few months.
This substantial improvement in the UV-transmittance must be maintained in the final application. Therefore solarization at the wavelength in use is generally not wanted. To test this ULTRAN glasses have been exposed to 365 nm light for 18 hours with a total dose of 2000 J/cm^2. The transmission before and after exposure shows no perceivable solarization.

1.5 Optical glasses with high homogeneity in diameters up to 1555 mm

For applications in astronomy, in wind tunnels, in laser research and for optical elements in interferometers one needs high homogen glasses in big sizes.
The most common glass, BK 7, has been successfully produced in sizes up to 1555 mm diameter and thickness of about 100 mm with a homogeneity of $\pm 1 \times 10^{-6}$.

In order to guarantee this high quality level overlapping interferometer photos are made with our 24 inch interferometer. The picture 6 shows this BK 7 disc during the interferometer test.

1.6 Highly prestressed windows for bubble chambers

Bubble chambers are detectors for particles produced in high energy particle accelerators. In a bubble chamber filled with liquid hydrogen one can make photos of the track of an electrical charged particle; the track consists of small gaseous bubbles arising behind the way of the particle.
The window needed must withstand the pressure of 20 bar and the low temperature of liquid hydrogen (T = 20°K).
Therefore it is necessary to prestress it.

SCHOTT supplied a BK 7 disc of diameter of 930 mm and a thickness of 173 mm, which received a special heat treatment to introduce a compression of 600 nm/cm. This was verified with birefringence measurements in axial and radial directions see picture 7 and 8.

Special events will be analysed exactly with computer aid.

Therefore a tight specification limits also the ray deviation through the window; this has been measured with an auto collimation telescope and documented.

2. Radiation Resistant Glasses

They are used as optical components in nuclear reactors and in nuclear isotope laboratories (hot cells),especially for remote observation instruments (telescopes) and remotely operated microscopes and television cameras. The radiation field of outer space is also an area of application for these glasses.

Radiation resistant optical glasses are needed if the total anticipated dose exceeds 10 Gy or the dose rate is greater than 0.05 Gy/h.

The addition of cerium to the composition allows optical glasses to be highly stabilized against discolouring due to radiation. After a dose of 10^4 Gy they only exhibit a slight discolouration, which barely increases if the dose is increased to 10^5 and 10^6 Gy.

Cerium is an electron acceptor which removes the radiation created free electrons in the glass matrix. Several glasses are stabilized to a great extent by large additions of cerium; they are especially suitable for use under conditions of extremely high radiation doses. Other glass types are only slightly stabilized.

SCHOTT offers in total 32 stabilized optical glasses, which are equally distributed over the range of glasses, see figure 10. Even some glasses for apochromatic colour corrections like FK 52 G 12 and KzFS 4 G 20 have been stabilized also. All glasses have been tested against ionizing radiation of electrons, protons, neutrons and gamma's.

They all show only a slight change in transmission even after a dose of 10^5 to 10^6 Gy.

3. Special glasses for the use in Cerenkov and Scintillation counters for particle research

Cerenkov Counters:

Cerenkov counters are detectors. They serve both for the identification as well as the determination of the velocity and charge of fast particles.

If the velocity of these fast particles when passing through a glass block radiator is greater than the phase velocity of light in this material then the characteristic blue Cerenkov radiation is emitted. The direction of propagation of this radiation is dependent on the velocity of the particle.

A light flash of less than 10^{-10} sec duration can be transformed into an amplified electronic signal and registered in a computer memory.

Demands on the Radiators:

Several highly refractive lead silicate glasses have proven to be especially good radiators. The following demands are placed on them:

- A high density of the glass brakes the fast particles in short distances. The radiation length is defined by the distance in which the energy of a particle is reduced to 1/e of its initial value. The higher the density, the smaller the glass block which can be chosen.
- A high refractive index lowers the detection limit because the threshold energy is dependent on the phase velocity in glass.

From the wide range of optical glasses from SCHOTT, several lead silicate glasses are particularly suitable for Cerenkov counters. See figure 11.

High internal transmission in the wavelength region from 400 nm to 480 nm is required in order not to dampen the weak Cerenkov radiation. Most highly refractive lead silicate glasses have a natural absorption in short wave spectral regions. See figure 12.

SCHOTT supplied more than 10000 blocks of Cerenkov glass for the OPAL detector installed at LEP (Large Electron Positron accelerator). This huge accelerator with a circumference of 27 km went into operation at CERN, Geneve.

Figure 13 shows part of the OPAL magnet and lead glass barrel during construction.

Scintillator glass HED-1

The advantages of the new scintillator glass in regard to the Cerenkov glasses are the high light emittance (about 25 times compared with SF 5) and the high energy resolution.

As glass can be made in relatively large volumes compared to crystals it is a more economic solution for a detector.

In addition it is more resistant against humidity and temperature-changes and can be handled more easily. The HED-1 glass is an important new development for future particle detectors.

4. Nonlinear optical effects in glasses

The striving for multifunctionality, for example, the combination of LASER and nonlinear optical properties, is a characteristic trend in presentday optical materials technology.

4.1 High energy LASER radiation creates nonlinear effects in glass and other materials.

Glass can be destroyed by such a LASER beam, because the incoming electromagnetic wave creates not only a linear polarisation but also a certain nonlinearity. This can be expressed via the so called "nonlinear refractive index" (n2).

Because the intensity of a LASER beam is greatest in the middle and decreases to the edge the beam is focused.

(n2 middle > n2 edge).

If this concentrated energy surpasses the damage threshold of the glass it will be destroyed. The calculated n2 value is increasing with higher refractive glasses. The n2 value is in the range of $0{,}5 \times 10^{-13}$ e.s.u. (FK 54) and 20×10^{-13} e.s.u. (SF 59). See figure 14.

4.2 Nonlinear effects in semiconductor doped glasses

Heterogeneous glasses consisting of uniformely distributed crystals of the type $CdS_xSe_{(1-x)}$ show higher nonlinear properties. After the melting process these crystals are created in the glass by a special heat treatment.

Depending on the CdS/Se content and the size of the crystals these glasses exhibit sharp absorption edges in the wave-length range between 400 and 850 nm. Therefore they are used as optical filter glasses. If the heat treatment is not uniformely made for example if a glass piece is heated with a temperature gradient, one can create a wedge filter.

Under the influence of a high energy LASER beam the absorption edge is shifted to smaller wavelengths.

This effect is reversible. One can use it for switching of light, like q-switching of a LASER or for modulation technique in telecommunication.

The wide range in observed response time (from nanosecond to femtosecond) led researchers to more detailed work to analyse nonlinear mechanisms.

CONCLUSION

The use of glass for optics and optoelectronics is determined by our high capabilities to vary the properties of materials in the vitreous state.

Glass offers new developments to satisfy the ever more complicated requirements of designers. Novel glass processing techniques and novel glass compositions are being developed for advanced applications.

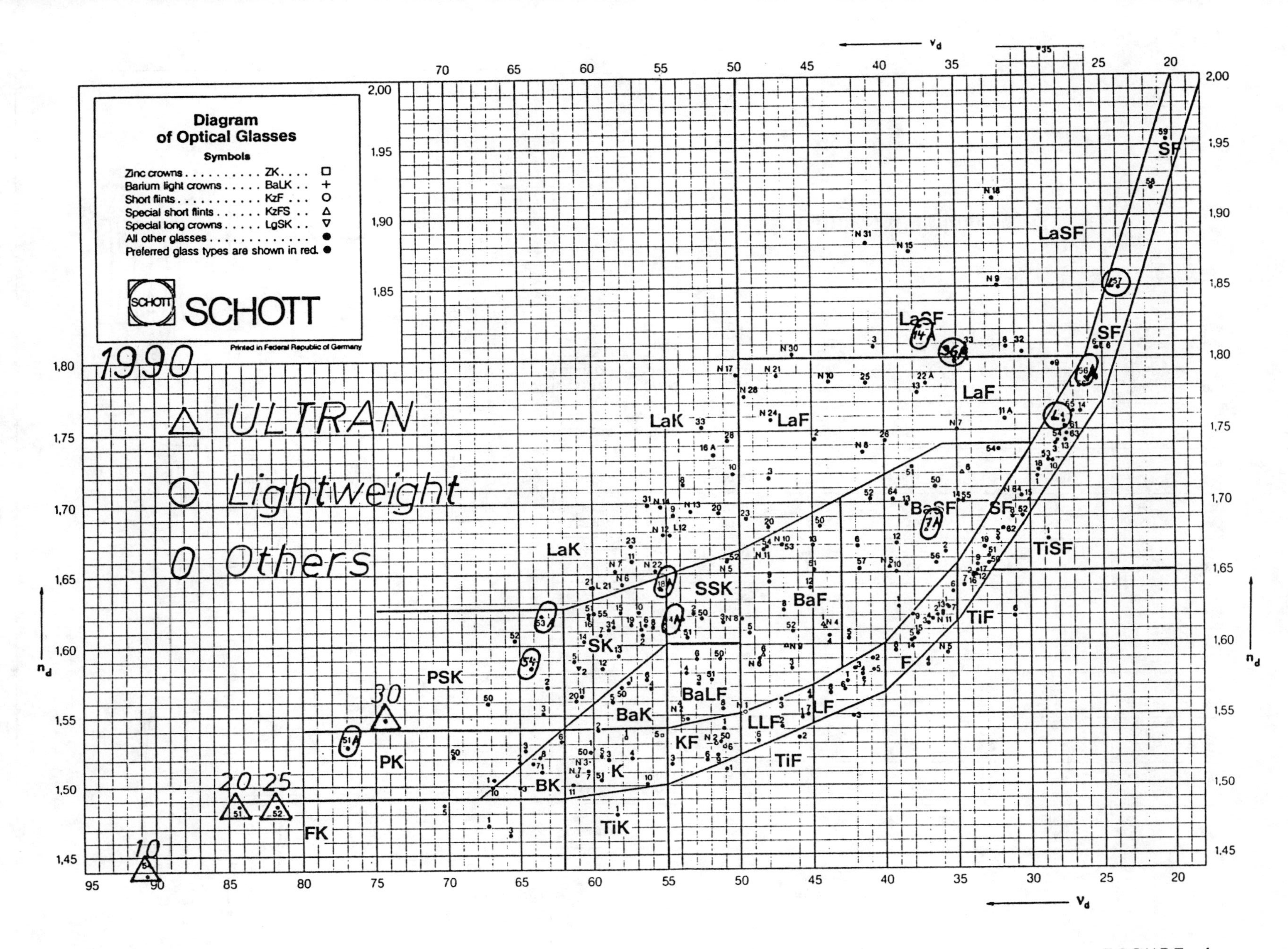
Diagram of Optical Glasses
Symbols
Zinc crowns . . . . ZK . . . .
Barium light crowns . . . . . BaLK . .
Short flints . . . . KzF . . .
Special short flints . . . . . . KzFS . .
Special long crowns . . . . . LgSK . .
All other glasses . . . .
Preferred glass types are shown in red.
SCHOTT
Printed in Federal Republic of Germany
1990
ULTRAN
Lightweight
Others
ν_d
n_d
SF
LaSF
TiSF
LaF
BaSF
TiF
F
BaF
LF
LLF
LaK
SSK
BaLF
KF
BaK
SK
K
TiK
PSK
BK
PK
FK

FIGURE 1

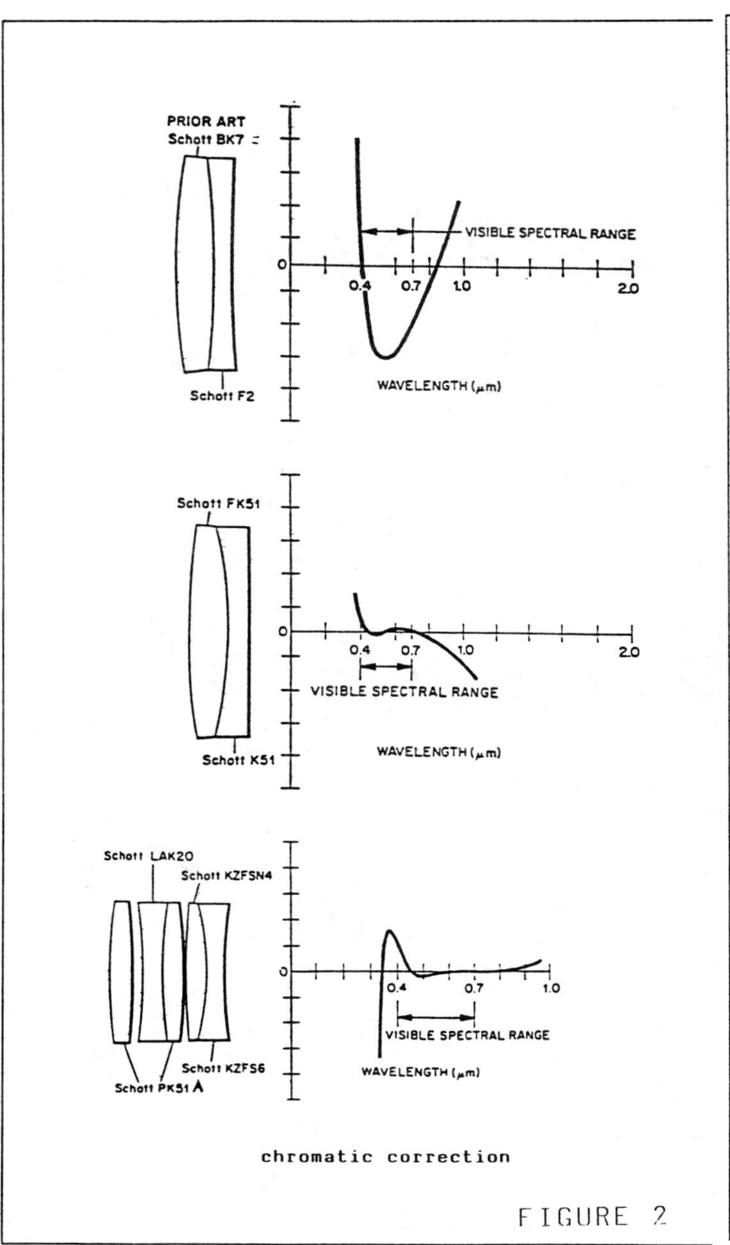

chromatic correction

FIGURE 2

Glasstype	n_d	ν_d	ρ (g/cm³)	weight reduction %	SR	AR	Tg (°C)	HK
F N11	1,62096	36,18	2,66	26	1	1.0	572	510
F 2	1,62004	36,37	3,61		1	2.3	432	370
SF L4	1,75520	27,21	3,37	30	1.0	1.3	569	500
SF 4	1,75520	27,58	4,79		51	2.3	420	330
SF L6	1,80518	25,39	3,37	35	2	1	586	500
SF 6	1,80518	25,43	5,18		52	2.3	423	310
SF L56	1,78470	26,08	3,28	33	2	1.3	591	530
SF 56A	1,78470	26,08	4,92		3	2.2	429	330
SF L57	1,84666	23,75	3,55	36	1	1	598	510
SF 57	1,84666	23,83	5,51		52	2.3	422	300
SF N64	1,70585	30,30	3,00	26	1	1.2	578	500
SF 15	1,69895	30,07	4,06		1	1.2	455	370
BaSF 64A	1,70400	39,38	3,20	19	2	1.2	580	540
BaSF 13	1,69761	38,57	3,97		51	1.2	584	450
LaSF 32	1,80349	30,40	3,52	28	1-2	1.0	544	560
LaSF 8	1,80741	31,61	4,87		52	1.2	476	420
LaSF 36A	1,79712	35,08	3,60	20	3	1.0	640	570
LaSF 33	1,80596	34,24	4,48		51	1.2	543	440
LaK L12	1,67790	54,93	3,32	19	52	2.2	636	600
LaK N12	1,67790	55,20	4,10		53	4.2	621	470
LaK L21	1,64048	59,75	2,97	21	53	4.2	641	590
LaK 21	1,64050	60,10	3,74		53	4.2	627	460

Lightweight glasses

FIGURE 3

SCHOTT

IMPROVEMENT
of internal transmittance
(thickness = 25 mm)

	400 nm	365 nm
FK 54	0,988 (0,98)	0,981 (0,96)
FK 52	0,992 (0,97)	0,961 (0,87)
BaK 1	0,993 (0,976)	0,975 (0,93)
SK 16	0,978 (0,97)	0,891 (0,81)
KzFS N4	0,972 (0,94)	0,872 (0,78)
KzFS N5	0,935 (0,90)	0,638 (0,49)
LaSF N30	0,924 (0,89)	0,662 (0,61)
LaSF N15	0,842 (0,71)	0,469 (0,27)
LaSF N31	0,824 (0,75)	0,452 (0,38)

( ) : catalogue values from 1980.

FIGURE 4

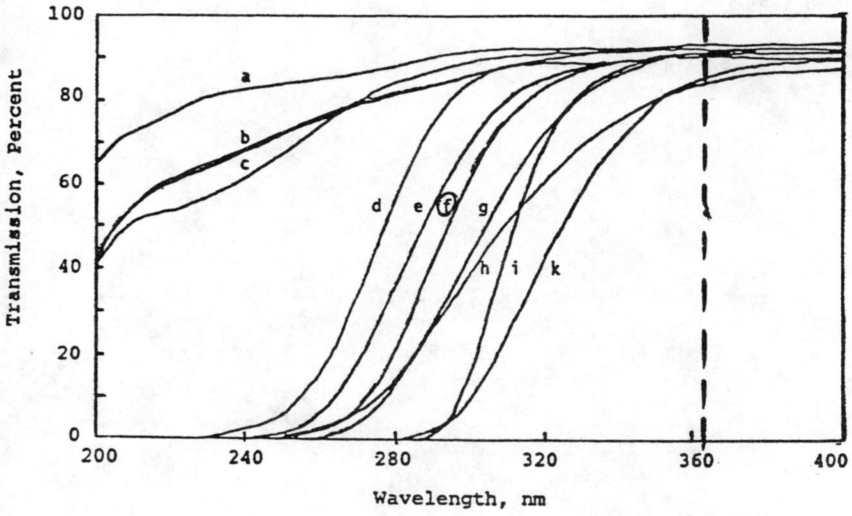

Ultraviolet transmission spectra of various glasses (5 mm thickness) a.) Ultran 10, b.) Ultran 30, c.) Ultran 20, d.) BK-10, e.) BaK-5 f.) UBK-7, g.) FK-3, h.) LaK-8, i.) PK-2 and k.) SK-16

FIGURE 5

SCHOTT

FIGURE 6

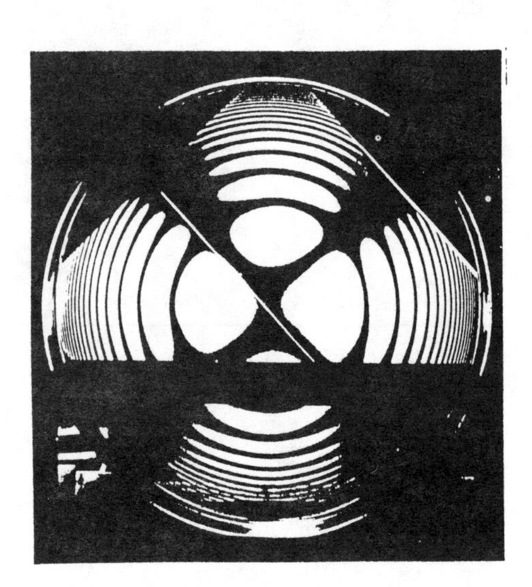

axial direction

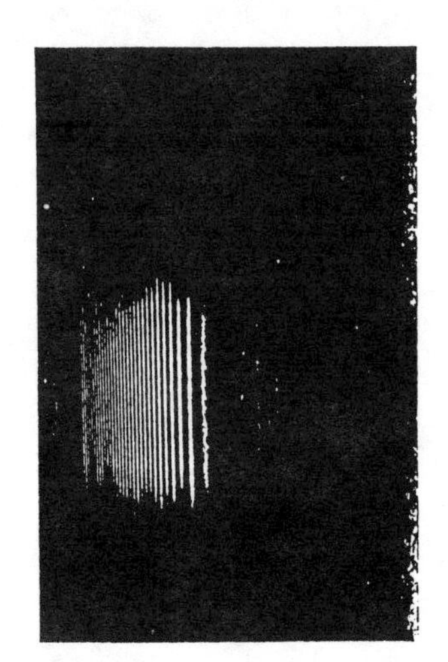

radial direction

stress optical measurement

FIGURE 7

Glass type	Refractive index n_g (435.8 nm)	Density (g/cm³)	Radiation length l (cm)	Critical energy* E_0 (MEV)
F 2	1.64202	3.61	3.224	19.90
SF 2	1.67249	3.86	2.804	17.95
SF 5	1.69985	4.07	2.551	16.90
SF 1	1.74916	4.46	2.166	15.17
SF 3	1.77444	4.64	2.037	14.66
SF 4	1.79121	4.79	1.907	13.85
SF 6	1.84705	5.18	1.698	13.09
SF 57	1.89391	5.51	1.546	12.44

* The critical energy is calculated from the proportions of the critical energies of the individual elements used in the glass. It is defined by the electron energy E_0, which is used in collision losses per radiation length.

Glasses for Cerenkov Counters

FIGURE 9

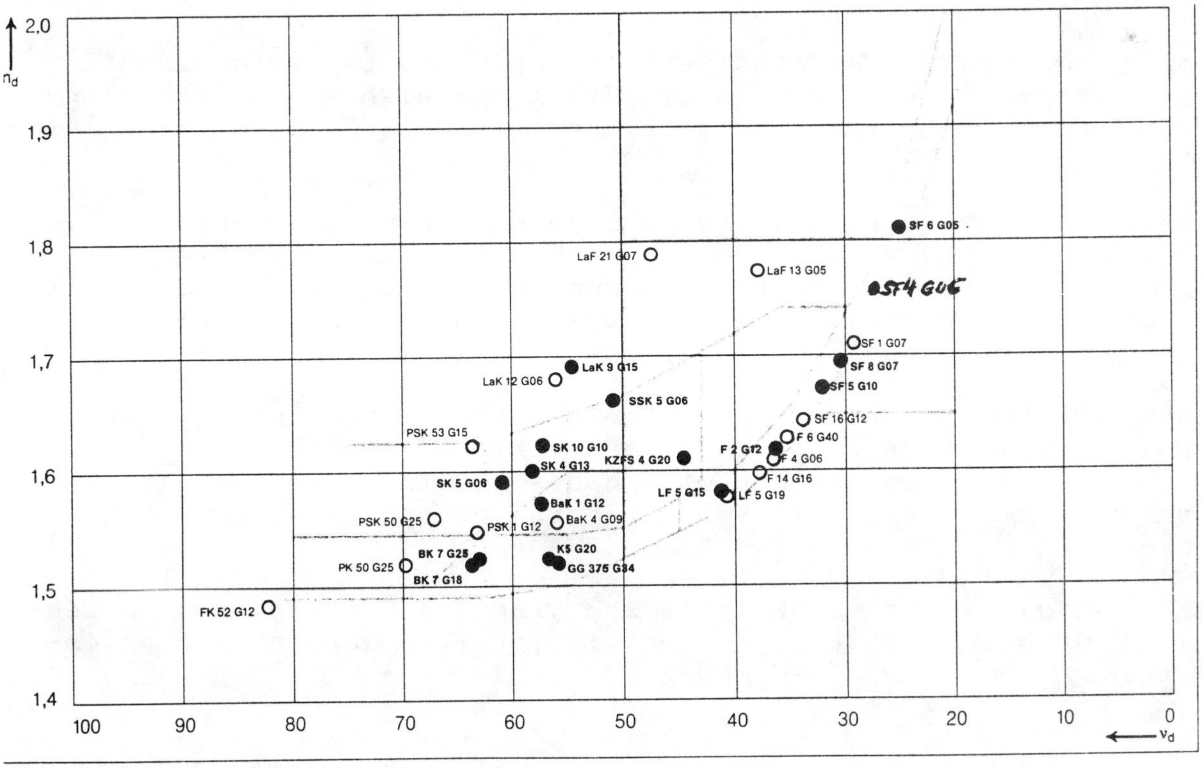

The optical position of the radiation resistant optical glasses.
● Available glasses, O Inquiry glasses.

FIGURE 8

High speed oscillation free lapping and polishing process for optical lenses

G. Richter

Wilhelm Loh Wetzlar, Optikmaschinen GmbH & Co. KG
D - 6330 Wetzlar / Germany

Abstract

LOH created a method of oscillation free lapping and polishing of optical lenses which led to a considerable increase of quality.

This method, named "SYNCHROSPEED", consists of computer calculations for tools and machine set up data, new designed tools and workpiece-holders and also specially developed machines.

These facts enable the user to achieve a high repeatability even for difficult lenses. Last not least the set-up and production times were drastically reduced.

1. Introduction

The manufacture of optical elements is characterized by precision and tolerances, with significantly higher accuracies than known in metal work.

Although there have been considerable improvements in the last 50 years, the limits were reached employing the high speed technology for lens surfaces better than 3 fringes with half fringe irregularity.

The problem of the tool and radius changes during the generating, lapping and polishing operations are a well-known fact. Frequent measurement of the lenses, time-consuming resettings of the machines, corrections of the diamond, lapping and polishing tools were expensive and unproductive activities.

In consideration of these facts, there was an urgent need to develop a process which would guarantee a high degree of reproducibility in the individual stages of processing, thus permitting automatic production using unskilled operators instead of highly skilled opticians.

After long and intensive analysis and practical trials, LOH succeeded in developing a high speed oscillation free lapping and polishing technology, named "SYNCHROSPEED".

This name was chosen to focus on a significant factor of the process. The relative speed of opposite points on the contact surface is synchronized by the mathematically fixed ratios of the angular speed. After developing a solid technological basis, the translation of theory into practice followed. The existing machine designs could not be used for this new manufacturing technology. Due to this fact, new machines, tools, lens holders etc. were created.

Additional feature included:

- Specific planning of production output through the use of computer-aided planning (CAP)
- Tool design to specified machinery parameters
- Lens holders for lapping and polishing, cementless and distortion free
- New adaption method tool for workpiece holders
- Automatic workpiece loading system
- Modular design
- Machine settings according to computer calculated data

The following shows in brief how these goals were achieved.

2. The "SYNCHROSPEED" computer software

The programme works in dialog with the user and covers the computer calculation of:

- Single surfaces
- Concentric surfaces etc.
- Geometrical shape of the tools
- Tool manufacture data
- Number and size of diamond-pellets, incl. allocation plan
- Shape of polishing foil
- Compensation data for tool wear
- Machine set-up data etc.

The calculated data is printed out or plotted. The programme itself is available on floppy disk and runs on IBM-compatible PC's.

All parameters necessary for tool manufacturing and machine settings can be analytically determined in advance during process planning.

These general considerations are now followed by some concrete examples:

3. Tools for the "SYNCHROSPEED" process

In the following the required tools and processing steps are explained for a single lens.

Fig. 1 shows various forms of generating tools. Types 1 and 2 have sintered rings; Type 3 is fitted with half diamond pellets and as Typ 2 provided with the workpiece radius.

The workpiece spindle can be moved 70 mm vertically. The grinding spindle feed during the cycle is in the direction of the tool axis.

Fig. 2 shows a tool set for the convex surface of the lens. The necessary data for the lapping tools are supplied by the computer printout (Fig. 3). The data to compensate the wear of the diamond-pellets are also calculated (Fig. 4).

The next picture shows the lens during lapping. The lens is pressed against the tool by the air pressure operated rubber membrane (Fig. 5).

The computer also calculates the data for the polishing tool and plots the shape of the polishing foil (Fig. 6). Other foil shapes are shown by Fig. 7.

4. Special machinery for the "SYNCHROSPEED" process

As mentioned before, completely new machinery types had to be developed for the "SYNCHROSPEED" process, and this is the key to the success.

All generating machines of the Spheromatic range operate by the high speed grinding method and possess revolutionary feed systems resulting in increased performance and long tool life (Fig. 8).

On the one hand, the demanded high precision optical surfaces require stable and easily adjustable machines, on the other, the deletion of oscillation ensures a reduction in complexity which in turn opens possibilities for automatic loading and unloading.

The generating machines are based on a modular design. They were further developed to use the major parts also for lapping and polishing machines so that they could be integrated in the system. A main feature is the infinite adjustment of the workpiece spindle RPM with digital display so that the speed factor calculated by the computer can easily be set.

Three sizes in the Spheromatic / Synchrospeed machine range are available:

Size 1 - up to 40 mm workpiece diameter
Size 2 - 30 - 90 mm workpiece diameter
Size 3 - 80 - 300 mm workpiece diameter

Size 1 and 2 include an option for automatic loading system.

4.1 **Workpiece holders** (Fig. 9)

The specially developed lens holders consist of a basic adapter with a self locking taper for attaching to the workpiece spindle, a special rubber membrane and an external plastic support ring. There is a sealed space between the membrane and the basic adapter which can be evacuated or pressurized with air pressure. At the upper edge of the membrane, there is a sealing lip which acts like a suction cup and prevents the lens from rotating during the cycle.

During processing, the workpiece is pressed against the tool by the airpressure operated membrane. A precision gauge allows optimized setting of the grinding pressure. The flexibility of the membrane assures uniform contact between tool and workpiece. Lateral forces are absorbed by the plastic ring.

5. **Are there typical workpieces for the "SYNCHROSPEED" process?**

In principle, all lenses with an edge thickness of at least 1 mm and an included angle of not more than 90° may be processed. Particularly difficult lenses with a diameter/thickness ratio of over 20 : 1 and more may be produced economically without encountering springing problems.

To date, around 30 various (and difficult) optical glass types, as well as Cerodur and IR-materials were processed successfully.

The radius range, so far processed, is $\pm$ 14 mm to ∞, the diameter range is 11 - 300 mm.

Economical batch sizes for the "SYNCHROSPEED" process are as follows:

Lens diameter	up to 60 mm	more than 250 pcs.
Lens diameter	60 to 100 mm	more than 150 pcs.
Lens diameter	100 to 300 mm	more than 50 pcs.

With difficult lenses a batch size of only 20 elements can be economical. Examples are shown on Fig. 10 / 11 / 12.

6. Information regarding SYNCHROSPEED from three important customers

How long has SYNCHROSPEED been applied?

- Since 1983

Which types?

- SPM/SPS 35; SPM/SPS 100; SPM/SPS 300

Time for tool manufacture?

- 2 - 3 hours/set depending on size.
  Polishing tool: Apply foil and dress 30'.

Life-time of tools?

- Lapping: 2000 - 6000 surfaces, depending on lens type and diameter
  Polishing: 200 surfaces for unfavourable lenses
  up to 2000 surfaces for favourable lenses, depending on radius

How large is the level of reproducibility?

- For the 35 and 100 lines: 500 surfaces.
  For the 300 line: 20 - 50 surfaces. For accuracy 3/2 (0.4) to 3/3 (0.6) between 4 and 8 hours. In practice this is equivalent to 40 - 60 surfaces compared to 6 surfaces using the conventional method with correction.

Processing times?

	Generating	Lapping	Polishing
- 35 line:	20" - 40"	10" - 20"	2' - 4'.
100 line:	30" - 80"	15" - 40"	3' - 8'.
300 line:	40" - 120"	20" - 60"	4' - 12'.

Setting-up time / machine ?

- Generating: 10' - 20' with magazin tray
  Lapping: 5' - 8'
  Polishing: 5' - 8'

Amortization time?

- 2 - 3 years.

Yield

- Yield 90 - 95 %
  Ratio of surface accuracy errors:
  - 1-2 % with Synchrospeed
    5-7 % with conventional method

7. Summary

An important factor for the "SYNCHROSPEED" process is the uniform wear of the lapping and polishing tools. This avoids radius changes and, together with the calculated machine set-up data, reproducibility is ensured. The geometrical configuration of the lapping and polishing tools is computed by the PC and printed or plotted.
The precision and regularity of the polished surfaces achieved, could only be attained in the past by the use of pitch polishing tools and highly skilled opticians.

To conclude, the 15 significant advantages of the "SYNCHROSPEED" process again:

1. Calculation of all production data without empirical trials.
2. Machinery set-up to specified calculated values.
3. Unusually short set-up times.
4. Drastic reduction in manufacturing times.
5. Reproducible quality standard.
6. Production of single lenses without blocking, if edge thickness is 1 mm or more.
7. Simple production of difficult parts, such as lenses sensitive to deformations due to their diameter/thickness ratio (e.g. 20 : 1)
8. Can also be used for IR-materials.
9. Production of high-quality optics by semi-skilled staff; precision opticians needed for tool preparation only.
10. Economy with small batches from 50 surfaces. Even a 20 pce. batch size is worth the investment in the case of problematic lenses.
11. Machine loading: manual, with loading device or fully automatic.
12. Ideal for finishing work, polishing off oxide layers, repolishing already centred lenses because the centering is not affected.
13. Higher yield of production.
14. Integrated workplaces, i.e. increased efficiency, short deliveries (Fig. 13).
15. Short-term amortization.

8. Reference

E. Brück; Das Synchrospeed-Verfahren.

Optical properties of Li-doped ZnO films.

A.Valentini, F.Quaranta, L.Vasanelli[1], R.Piccolo

Department of Physic, University of Bari
Via Amendola 173, 70126 BARI (Italy)
[1]Department of Material Sciences, University of Lecce
Via per Arnesano, 73100 LECCE (Italy)

ABSTRACT

The difficulty to achieve a refractive index matching between active substrate and active layer grown on, is one of the main problem in integrated optical devices based on gallium arsenide, because of its high refractive index value. One possible solution could be an active layer whose refractive index is variable during the grown. Zinc oxide is a very intresting material because of its electro-optic and acusto-optic properties. It has a low cost and can be prepared by a variety of techniques. In this paper deposition of lithium doped zinc oxide films by reactive sputtering has been investigated in order to study the dependence of optical properties on lithium content and deposition parameters. A ZnO:Li target was used. The film depositions were performed varying the oxygen content in sputtering gas. For comparison undoped ZnO films were also prepared. We have performed optical and electrical measurement on films relating the results to Li contents and O/Zn ratio obtained by nuclear reaction and Rutherford backscattering measurements respectively. The film analysis has shown that dopant concentration is mainly controlled by gas mixture. The optical properties are dependent on deposition conditions. Optical waveguides have been prepared and characterized. The results are presented and discussed.

1.INTRODUCTION

With the progress of integrated optical devices on gallium arsenide, there is an increasing interest in developing active layers. Because of substrate high refractive index one possible solution could be an active layer whose refractive index can be varied during the grown.

ZnO film is an excellent piezoelectric material extensively used in acoustic surface wave as well as in acoustic-photon devices.

This paper reports on the preparation of lithium doped zinc oxide films by reactive radio frequency sputtering. The material properties have been investigated in order to study the dependence of the optical behavior on lithium content.

EXPERIMENTAL DETAILS

The deposition apparatus, consisting of an Alcatel r.f. diode sputtering, is shown in Fig.1. The target used is a sintered disc, 10 cm-diam, of ZnO:Li (1mol% Li/ZnO ratio). The chamber was pumped to a residual pressure of $5x10^{-7}$mbar before the sputtering gases, 5N argon and 5N oxygen, were inflated in various ratios. The gases were supplied to the deposition system through a MKS digital controller gas flow system. The pressure was kept constant at the value of 2.5×10^{-3}mbar. The substrate used were Corning glass 7059, for optical and electrical measurement, and carbon, for elemental composition analyses. The substrate holder was maintained at 300 °C during deposition by a Eurotherm temperature controller. The r.f. forward power density was 5.05 W/cm^2. The deposition time was 3600 s. For comparison some samples of undoped ZnO films have been prepared in the same growing condition starting to a 5N purity ZnO target.

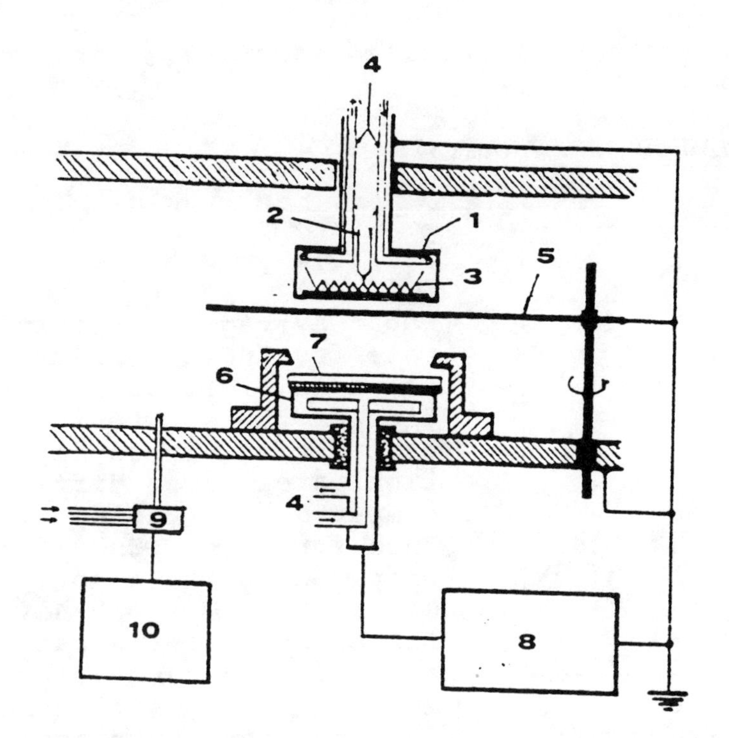

Fig.1 Schematic diagram of rf diode sputtering apparatus: 1.Substrate holder; 2.Thermocouple; 3.Heater; 4.Water cooling; 5.Shutter; 6.Target holder; 7.Target; 8.R.f. generator; 9.Flow meter; 10.Flow control unit.

The average deposition rate was determined by measuring the film thickness with Alphastep 200 profilometer and dividing the value of the thickness by the length of time taken to make the deposition.

In order to obtain information on the O/Zn ratio Rutherford backscattering (RBS) was carried out. Because of the small atomic mass of Li, nuclear reaction analysis (NRA) was performed to investigate the dopant concentration. Details on the apparatus and the measurement procedures are reported elsewhere[1].

The x-ray diffraction peak profiles were recorded with a Philips powder diffractometer having Bragg-Brentano geometry. CuKα radiation was used. The electrical properties were obtained by measuring at STP the film electrical resistivity with Van der Pauw method.

Transmittance measurements have been carried out to study the optical behavior of the film in the spectral range 0.3-2.5 um using a Perkin-Elmer 330 model spectrophotometer.

Some $10x15mm^2$ slabs of Corning 7059 glass have been used as substrate to obtain optical waveguide. A prism coupling technique, using rutile prism, was used to characterize them. A detailed description of the measurement is reported elsewhere[2].

3. RESULTS AND ANALYSIS

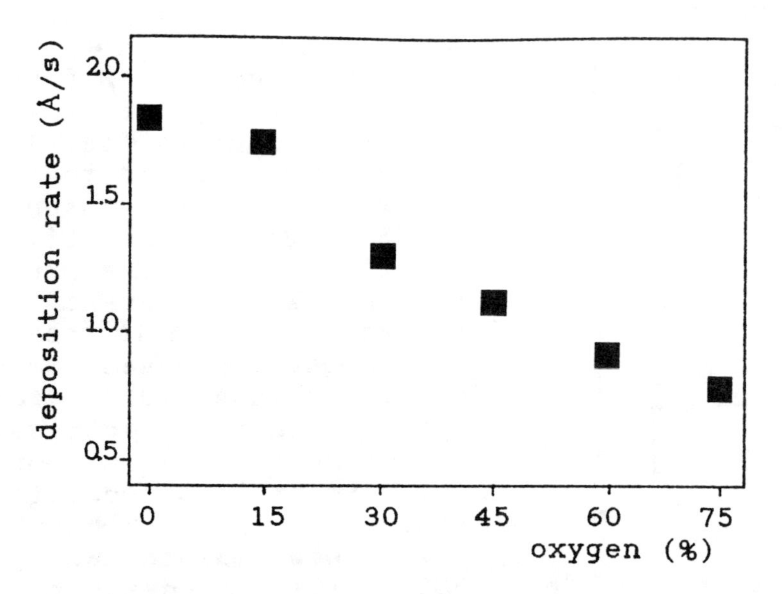

Fig.2 Average deposition rate versus oxygen percentage in the sputtering gas.

3.1 Film growth rate.

The average deposition rate for ZnO:Li films is shown in Fig.2 as a function of the sputtering gas composition. These deposition rates are different compared with those obtained for pure ZnO films prepared in the same condition[3,4]. The behavior for Li doped films can be explained in term of reduction of deposition rate due to the increase of light ion percentage in the sputtering gas[5].

3.2 Nuclear analysis

The O/Zn ratio and the Li concentration in the doped ZnO films are resumed in Table I. The same ratio measured for undoped films is reported for comparison. In order to obtain the Li concentration a uni-

Table I. Summary of sputtering gas percentages and film composition obtained by nuclear measurements.

Sample Ident.	Target Comp.	Gas(%) Ar	Gas(%) O_2	O/Zn ratio (±0.01)	Li/Zn mol% (±8%)
391	ZnO	100	---	1.04	----
392	ZnO	85	15	1.03	----
393	ZnO	70	30	1.01	----
394	ZnO	65	45	1.01	----
395	ZnO	40	60	1.02	----
396	ZnO	25	75	1.03	----
398	ZnO:Li	100	---	1.03	1.0
399	ZnO:Li	85	15	1.04	1.7
400	ZnO:Li	70	30	1.03	2.7
401	ZnO:Li	65	45	1.03	2.9
402	ZnO:Li	40	60	1.07	3.4
403	ZnO:Li	25	75	1.11	4.3

form distribution of dopant in the film was assumed[6].

The sample grown in pure argon have the same lithium concentration of the target. The oxygen introduction in the sputtering gas produce an increase of the Li concentration in the films. Only for heavily doped films the lithium presence affects the stoichiometry. The increased oxygen incorporation can be explained by the possible formation of lithium oxide compounds in these films. This is also suggested by the absence of stoichiometry change in undoped films grown varying the oxygen percentage in the sputtering gas.

3.3 Structural and elecrical analyses

The x-ray diffraction analysis has shown that Li doped ZnO films are oriented with the c-axis normal to the substrate. The x-ray peak position shift is of the same order of that found in undoped ZnO films[7,8]. This means that the Li introduction does not yield further stress in the film[9].

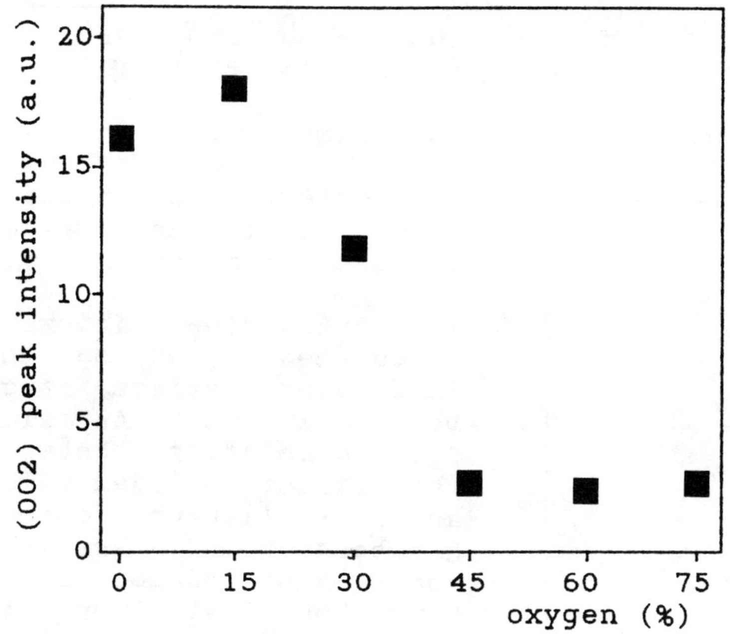

Fig.3 (002) peak normalized intensities versus oxygen percentage in the sputtering gas

In previous work[1] we have shown that the Li introduction gives rise to a lowering in the ZnO film preferential orientation. The (002) peak normalized heights reported in fig.3 confirm those results. The curve shows a maximum due to the improvement of the film structural quality dependent upon the O_2 content of the sputtering gas, similar to that reported for undoped ZnO films[3,4]. The shift of the maximum towards lower oxygen percentage can be ascribed to the dopant introduction in the film produced by the oxygen.

Regardless dopant concentration all the doped films have shown electrical resistivity value of about 10^7 Ωcm. These values are lightly less than those achieved for pure ZnO films, but suitable for acusto-optical application.

3.4 Optical analysis

A typical transmittance spectrum obtained for Li doped film is shown in fig.4. All the films exhibit a transmittance higher than 70% in the 0.4-2.5 um spectral range causing the film clear appearance. The

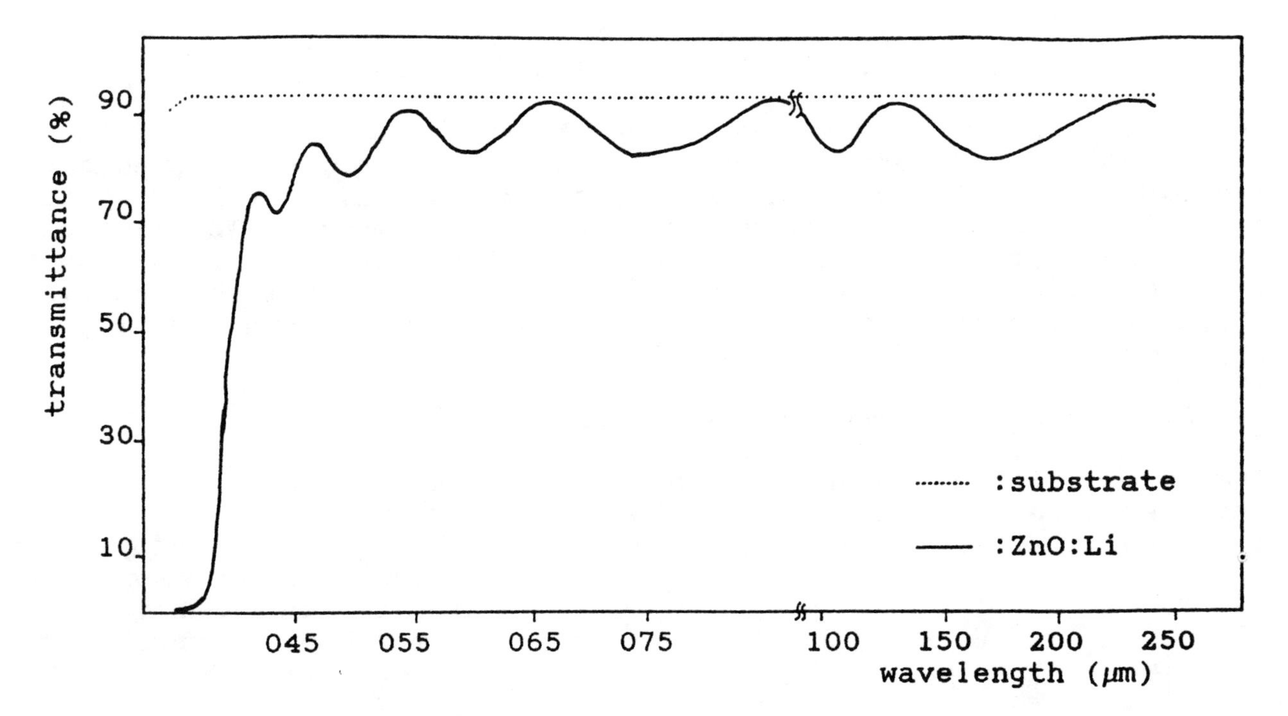

Fig.4 Typical transmittance spectra for ZnO:Li films.

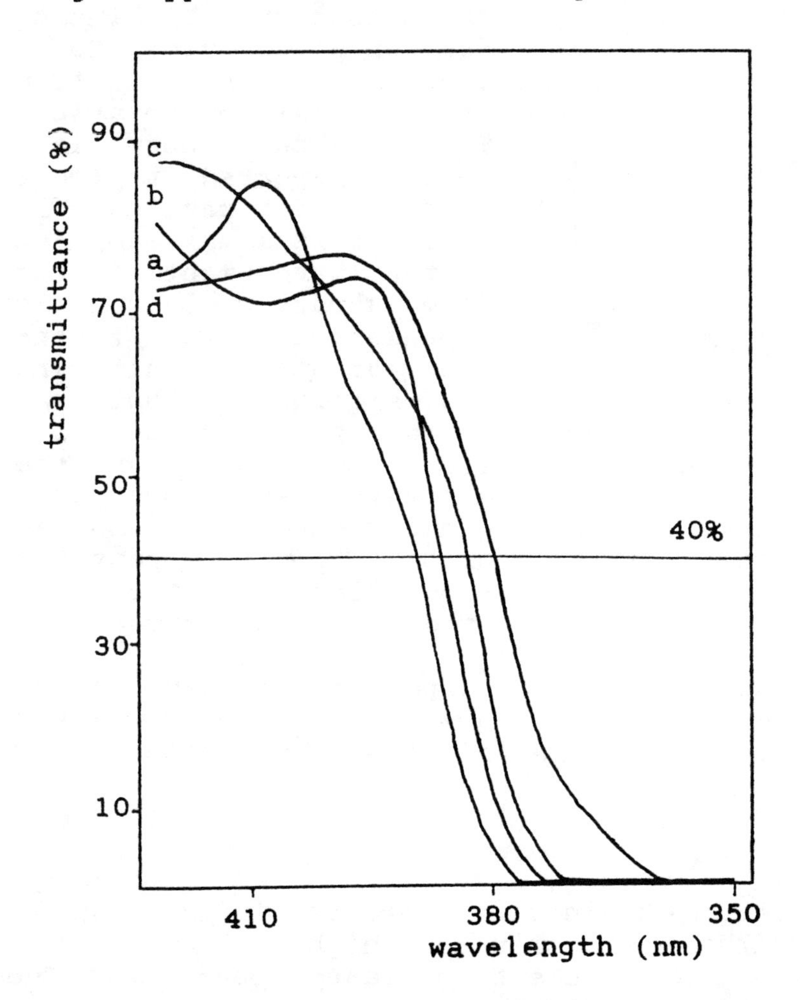

well determinate interference minima and maxima indicate film uniformity.

The refractive indexes, calculated by using the interference effect, range between 1.85-2.1. Any simple correlation between the refractive index value and the lithium content has been found. Previous analyses on undoped films have shown that there is an increment of the refractive index value when oxygen is added in the sputtering gas[1,10]. For the doped films this effect is in contrast with the structural uniformity worsening due to lithium in-

Fig.5 Transmittance spectra near the absorption edge versus Li/ZnO mol% : a. undoped; b. 1.0 mol%; c. 2.7 mol%; d. 4.3 mol%.

troduction, enhanced by the oxygen.

In fig.5 are reported some transmittance spectra of doped and undoped samples near the absorption edge. A simple evaluation of the optical gap by the transmittance cutoff shows a shift in the optical gap towards higher energy for all the doped films. Moreover it can be observed that this shift is enhanced by the lithium concentration in the films. This shift can be probably ascribed to the formation of lithium oxide compounds that have the optical gap in UV region[11] at energy higher than undoped zinc oxide.

The analyses of the optical waveguides have shown that was very hard to couple the light in the devices. Moreover the light scattering was high and it was not possible to measure the optical attenuation. This confirms the worsening in the crystalline structure due to the lithium presence.

4.CONCLUSION

Lithium doped ZnO films were grown by rf sputtering on amorphous substrates.
By choosing suitably the oxygen percentage in the sputtering gas it is possible to control the lithium incorporation in the film.
Lithium concentration higher than 3% is able to affect the film stoichiometry with an increase of the oxygen to zinc ratio.
The structural properties are considerably affected by the lithium incorporation and any improvement is observable even if the oxygen is introduced in the sputtering gas.
The ZnO film transparence is not affected by lithium introduction.
The more controllable optical effect due to lithium presence is a shift of the optical gap towards higher energy. The shift increase is proportional to the film lithium content.
Because of the structural worsening, further crystallization procedure, such as heat post deposition treatment, are probably needed in order to use lithium doped ZnO films in optical waveguides.

5.ACKNOWLEDGMENTS

The authors would like to thank Prof.C.Battaglin for the discussion about nuclear analyses. Thanks are also due to Mr.G.Casamassima for the technical support to produce the films. This work is partially supported by Progetto Finalizzato MADESS of CNR, Italy.

6.REFERENCES

1.A.Valentini, F.Quaranta, M.Rossi, C.Battaglin, *Preparation and Characterization of Li-doped ZnO Films,* to be published.

2.A. Valentini, F.Quaranta, L.Vasanelli, *Dielectric zinc oxide films characterization for optical waveguides,* Thin films in optics, T. Tschudi, 1125, 36-40, SPIE, Washington, 1989.

3.C.R.Aita, A.J.Purdes, R.J.Lad, P.D.Funkenbusch, *The effect of O_2 on reactively sputtered zinc oxide,* J.Appl.Phys., 51(10), 5533-5536, 1980.

4.M.Di Giulio, A.Valentini, L.Vasanelli, *Deposition of ZnO Films by R.F.Sputtering,* Mat.Chem.Phys., 9, 197-203, 1983.

5.G.K.Wehner, G.S.Anderson, *The Nature of Physical Sputtering,* Handbook of Thin Film Technology, L.I.Maissel, R.Glang, cap.3, 1-38, McGraw-Hill, New York, 1970

6.L.B.Wen, Y.W.Huang, S.B.Li, *The measurement of the trace Li content and distribution in Zno films and the thickness of these films,* J.Appl.Phys. 62(6), 2295-2297, 1987

7.R.J.Lad, P.D.Funkenbusch, C.R.Aita, *Postdeposition annealing behavior of rf sputtered ZnO films,* J.Vac.Sci.Technol., 17(4), 808-811, 1980.

8.A.Valentini, F.Quaranta, M.Rossi, L.Vasanelli, *Structural properties of ZnO films prepared by rf sputtering for optical applications,* Thin Solid Films, 175, 255-259, 1989.

9.S.Sen, S.K.Halder, S.P.S.Gupta, *An x-ray line profile analysis in vacuum-evaporated silver films,* J.Phys.D:Appl.Phys., 8, 1709-1721, 1975.

10.A.Valentini, A.Quirini, L.Vasanelli, *Optical properties of ZnO films deposited by rf sputtering on sapphire substrate,* Thin Solid Films 176, 1167-1171, 1989

11.K.Uchida, K.Noda, T.Tanifuji, Sh.Nasu, T.Kirihara, A.Kikuchi, *Optical absorption spectra of neutron-irradiated Li_2O,* Phys.Stat.Sol. (A), 58, 557-566, 1980.

APPLICATION AND MACHINING OF "ZERODUR" FOR OPTICAL PURPOSES

Norbert Reisert

SCHOTT GLASWERKE, Mainz/FRG

ABSTRACT

"Zerodur" is a glass ceramic made by SCHOTT GLASWERKE, exhibiting special physical properties, while also being optimally suited for a variety of applications.

Thermal expansion of "Zerodur" is zero over a large temperature range and temperature variations, thus, have no bearing on the geometry of workpieces, which makes "Zerodur" ideally suited for the use as mirror substrate blanks for astronomical telescopes, x-ray telescopes, or even for chips production, where maximum precision is a prime requirement. The temperature-independent base blocks of ring laser gyroscopes, as well as range spacers in laser resonators are likewise made of "Zerodur".

"Zerodur" can be machined like glass, but unlike with many optical glasses the warming generated upon cementing and polishing does not cause any deformations or tension at the surface.

The paper aims to provide a general view of the most essential proporties of "Zerodur", its major fields of application, the manufacture and the machining in the form of grinding an polishing.

APPLICATION AND MACHINING OF "ZERODUR" FOR OPTICAL PURPOSES

1. INTRODUCTION

The Schott Glassworks in Mainz, Germany, is Europe's largest manufacturer of special glass. Suited to comply with the most varied demands in application in optics, electronics, chemistry, astronomy, medicine, research, in household use, building, and in many other areas, there is a variety of glass types which are manufactured with greatly differing physical properties. In the Optics Division alone, a total of about 250 different optical glass types, about 100 coloured and filter glasses and about 50 different technical glasses are manufactured and processed. Examples of these are:

- glasses with a high refractive index and low specific weight for use in opthalmics

- glass systems with complementary properties for lenses for cameras, reproduction, equipment etc.

- coloured and filter glasses with transmission ranging from ultra violet to infrared

- lead glasses of high density for Cerenkov counters or radiation protection windows

- glasses with extremely low attenuation for light-leading fibres

- glasses with minimum expansion for fire protection

and last but not least

- "Zerodur", the glass ceramic with zero expansion for a multitude of uses which will be dealt with in detail in the following.

In the same way as there is a great variety of materials from which to choose, there is also a large variety of products, ranging from small lens and prism blanks for the optical industry through a whole range of intermediate forms and sizes to radiation protection windows with viewing areas of more than 1,000 x 1,000 mm or mirror blanks with a diameter of more than 4,000 mm for large astronomical telescopes. SCHOTT attempts at all times to comply with the wishes and requirements of the customers by supplying parts which are, to a great extent, preprocessed to exact measurements.

2. WHAT IS GLASS CERAMIC ?

Glass ceramic is an inorganic, poreless material which contains both glassy and crystalline phases. "Zerodur"'s crystalline phase amounts to about 70%. The mean size of the crystals is in a range of about 0.05 µm. As a result of the small size and the minor difference in refraction in comparison to the glassy phase, "Zerodur" possesses good transparency.

3. SPECIAL PROPERTIES OF "ZERODUR"

Negligible thermal expansion

The negative expansion of the crystalline phase counteracts, to the greatest extent, the positive expansion of the glass phase. The linear coefficient of thermal expansion α (0°C, 50°C) amounts to $0 \pm 0.02 \times 10^{-6}$/K and $0 \pm 0.05 \times 10^{-6}$/K respectively, according to class of expansion (in comparison, borosilicate glass has a figure of 3.2×10^{-6}/K).

Homogenity

"Zerodur" is extremely homogeneous not only in its thermal and mechanical properties but also in its optical properties. For example, in the case of a round disk of Ø 1,450 x 150 mm, a maximum variation of refractive index of

$\leq \pm 2 \times 10^{-6}$ was measured interferometrically, corresponding to a homogenity stage of H 3.

Chronological dimensional stability

Measurements over a period of 17 years show the exceptional chronological linear stability of "Zerodur".

The linear aging coefficient, with which the linear alteration can be measured, amounts to $< 1 \times 10^{-7}$/year.

4. MANUFACTURING PROCESS

"Zerodur" in large technical production is produced in a melting tank in the same manner as glass. In the case of smaller articles, this is carried out in a continuous melting process while large parts, such as telescope blanks are manufactured discontinuously with volumes of up to 50 tons.

Processing is also carried out in the same way as for glass by pouring into molds, drawing, rolling or pressing.

Determining factor for the total manufacturing process is that the glass melt exhibits sufficient crystaline stability for this to remain glassy during the subsequent processing and cooling. The later transformation to a partially crystalline state is made possible by the differing dependency upon temperature of the nucleation frequency and speed of growth of the crystals.

The nucleation maximum is at a lower temperature (approx. 770°C) than the maximum of the speed of growth (approx. 1,100°C).

Correspondingly, in spite of the high rate of growth, the glass cannot crystallize during annealing after casting as, apart from a few external nucleating agents, no crystal forming nucleating agents are present. On renewed passage through the nucleation range, a sufficient number of nucleating agents are formed enabling ceramizing crystals to grow.

This correspondingly places considerable demands on the coarse annealing:

- Monotonously decreasing cooling to room temperature. Localized reheating can lead to premature ceramizing and to breakage.

- Avoidance of temporary stress during cooling. Decrease in thermal conduction must be compensated by a reduction in the cooling speed in order to avoid breakage during cooling. It is also of equal importance to have a good temperature distribution in the annealing furnace (gradient $\leq$ 5°C).

- Obtaining small permanent stress. The permanent stress must allow any coarse processing required for the glassy parts.

The coarse annealing is followed by a quality control and, if required, by coarse processing.

There then follows the decisive temperature treatment, the so-called ceramizing process. The glassy material is converted to a partially crystalized state at a temperature of about 800°C. This is the case of an exotherm process in which 250 KJ per kg of material are set free. In addition, there is a volume shrinkage of 3 ‰. By means of the temperature/time program, the crystallite size and distribution and, resultingly, the coefficient of expansion, can be controlled.

Following cooling to room temperature, a careful check of the internal quality and measurement of the physical properties are carried out. The material is then available for further processing.

5. PROCESSING

The processing of "Zerodur" is carried out in the same manner and using the same machines and tools as for optical glass. "Zerodur" can be ground, lapped and polished; it can also be drilled with throughput or blind bore holes and the manufacture of lightweight structures, e.g., by grinding or water jet cutting presents no problems. Glassy, flat disk plates can be slumped and ceramized in one process as meniscus or glassy, light-weight structures welded and ceramizid.

Parts are delivered, custom-made and, to a greater or lesser extent, pre-ground; fine grinding and, if required, polishing is carried out by the final processor.

Grinding

The tools used here are galvanic coated diamond tools (mainly in the case of special shaped tools) and tools with a bronze or plastic bonding. The selection of the grain size depends upon the customer's requirements; in general, tools of between D 251 (70/80 US mesh) and D 64 (230/270 US mesh) are used. The surface roughness attained in this manner is between 100 µm (D 251) and 20 µm (D 64), the micro crack depth is between 300 µm 251) and 20 µm.

We use cupwheel cutters as well as peripheral cutters, and also cutting blades and diamond-studded core drills.

In most cases a peripheral speed of 25 m/sec has proved to be an optimum value with respect to removal rate, tool lifetime and sharpness. The other parameters like infeed rate and tool specification don't differ from optical glass. This is of great importance for manufacturers with a broad range of materials.

Polishing

We do not normally carry out final polishing, We do, however, carry out coarse polishing of plane surface for quality check for internal defects. For this purpose, we use both plastic and felt tools and, as polishing agend, Ceriumoxide.

As a result of the hardness of "Zerodur", the polishing times are longer than in the case of many optical glasses. We have learned from customers, however, that excellent surface qualities can be achieved by polishing. In the case of the polished ROSAT x-ray telescope mirror from Carl Zeiss, Oberkochen, the resulting surface roughness amounts to less than 4 Angstroem, i.e., 0,0004 µm.

Polishing objects with precise requirements has considerable advantages as the material is not deformed by heat caused during processing and delays caused by temperature in interferometer measurements, which accompany the polishing process, are minimal.

6. Examples of application

Astronomical mirror telescopes

"Zerodur" is used as a material for mirror blanks in high capacity telescopes both on the earth in the fields of visible and infra red light and in space in the field of x-rays.

The most recent examples are the NTT of the ESO with its active, 3,6 m main reflector and the Galileo telescope of the same type of the University of Padua and the nested x-ray telescopes of ROSAT and AXAF.

At present, SCHOTT is beginning manufacture of four monolithic reflector supports with a diameter of 8.2 m for the Very Large Telescope for the ESO. When set in beam combination these four reflectors correspond to a main reflector of 16 m and are, correspondingly, larger than the 10 m Keck telescope which is compiled of 36 "Zerodur" segments.

"Zerodur", as a result of its negligible thermal expansion, its high young's modulus and good stability in regard to its shape, which, together with exceptional imaging quality and good processing properties and excellent chemical resistance, which is important for cleaning, vapor deposition and removal of metalic reflector coatings, is predestined for such use.

Light-weight mirrors

As already mentioned, "Zerodur" possesses exceptional properties as a mirror blank support. Even although considerable reduction in weight can be achieved, the stability of its shape remains.

2 methods of reducing weight are applied:

- By means of blind bore holes or undercut bore holes in different basic shapes (round, hexagonal, triangular, rectangular) which make it possible to achieve a reduction in weight of up to 60 %.

- Fusing a honeycomb structure to a front and rear plate makes it possible to achieve a weight reduction of more than 80 % as opposed to a massive part.

Laser gyroscopes

In modern navigation, we can no longer do without laser gyroscopes. These consist of a basic triangular or rectangular body with a ringshaped drilling through which the laser beam is projected. The most important fact for exact measurement is that the basic body retains its measurements precisely during changes in temperature. "Zerodur", as a result of its non-existent thermal expansion, is ideally suited as a material for the units.

Spacers in laser resonators

In gas lasers, the plasma is situated between two reflectors which constitute the resonator. The distance between the reflectors must be maintained with the greatest degree of exactitude (1 µm at a length of 7 m). The distance between these reflectors is, therefore, set by means of "Zerodur" rods.

Devices for chip manufacture in the semi-conductor industry

Both reflector systems for the projection of the conductor paths on the Si-wafers, the prisms used for deflecting beams and the high precision standard square for fixing wafers may not lose their exactitude under the influence of temperature changes and are, correspondingly, manufactured from "Zerodur".

Reference surfaces

As a result of the negligible thermal expansion, the stability of shape and the good processing properties, "Zerodur" is ideally suited for use in reference surfaces, e.g. in interferometrics.

Transmissive optics

The excellent transparency and the aforementioned optical homogeneity of "Zerodur" also make it possible to use this in transmission optics, e.g.,as reflectors for partial transmission in interferometers.

The checking of convex secondary reflectors in mirror telescopes can also be carried out from the rear, through the "Zerodur" blank. In this way, we can avoid the necessity of manufacturing an expensive counter reflector.

Theory and experiment as tools for assessing surface finish in the UV-vis. wavelength region.

Joakim Ingers and Laurent Thibaudeau
Department of Physics II
Royal Institute of Technology
S-100 44 Stockholm, Sweden

The agreement between theoretical and experimental data for light scattering from random rough surfaces has been investigated for several different wavelengths of the incident field. Theoretically the scattering from two-dimensional surface roughness has a λ^{-4} dependence whereas scattering from unidirectional surface roughness has a λ^{-3} dependence. We have investigated this wavelength dependence experimentally. Our results show that perturbation theory gives a good agreement with experimental data when the root-mean-square (rms) roughness of the surface does not exceed about one twentieth of the wavelength of the incident field. These results have been obtained for scattering from several different materials and wavelengths indicating that theoretical calculations can be used as tool for assessing the influence of light scattering from surface roughness on system performance. The use of experimental surface profile data allows us to predict the scattering from a surface with good accuracy without any prior assumptions about the surface roughness. We have not bee able to verify experimentally the theoretical wavelength dependence of the scattered intensity.

1. INTRODUCTION

The scattering of electromagnetic fields from surface roughness has received a lot of attention in the past few decades, both in the radar literature and in the optics literature. In this paper we try to build a bridge between theoretical calculations of the scattered field and experimental measurements. We focus on a few points. First we explore the possibility of using experimental surface profile data for predicting the scattering of an electromagnetic field by surface roughness in the UV-visible wavelength region. We will also make a brief comparison between different solutions to the scattering problem. Secondly we attempt to verify experimentally the predicted wavelength scaling of the scattered intensity.

2. THEORY

There have been a number of different solutions to the scattering problem.[1] Each of these has its own niche and the choice of solution is dictated by the type surface and the characteristics of the surface roughness. Generally solutions to the scattering of an electromagnetic field by a randomly rough two-dimensional surface (*i.e.* a surface that can be described by a function of two variables) can be divided into the following categories;

1) Scalar theory,[2]
2) Fresnel-Kirchhoff solutions,[3]
3) Rayleigh-Rice perturbation theory.[1]

The first of these is a simple solution which does not account for the vector nature of the electromagnetic field.

First-order Fresnel-Kirchhoff solutions apply to gently ondulating surfaces, that locally can be approximated with a plane tangent to the surface. Fresnel-Kirchhoff solutions have usually been applied to rougher surfaces with large surface features.

Rayleigh-Rice perturbation theory is valid for surfaces whose roughness is small compared to the wavelength of the incident field. The conditions on the surface slopes are less severe than for the Fresnel-Kirchhoff solution. A recent study has shown that Rayleigh-Rice solutions yield results which are equivalent to solutions based on the theoretically more rigourous extinction theorem.[4]

The surfaces that have been used for this study are essentially microrough and consequently we have chosen to work with Rayleigh-Rice perturbation theory. For two-dimensional surfaces perturbation solutions are generally only carried out to first order since higher-order solutions are very complex.

2.1 First-order perturbation theory

An expression for the normalized scattered power is given by[5]

$$\frac{1}{P_0}\langle\frac{dP}{d\Omega}\rangle = \frac{(2\pi)^2\,|1-\varepsilon_1|^2\cos^2\theta}{\lambda^4\cos\theta_0}[|p_2|^2+|s_2|^2]g(\vec{K}), \tag{1}$$

where λ is the wavelength of the incident field, θ is the polar angle of scattering, and θ_0 represents the polar angle of incidence. p_{2r} and s_{2r} denote the optical factors and $g(\vec{K})$ is the power-spectral-density (PSD) function. <..> denotes an ensemble average. Notice the λ^{-4} dependence of the scattered power. In the case of infinitely conducting surfaces Eq. (1) reduces to

$$\frac{1}{P_0}\langle\frac{dP}{d\Omega}\rangle = \frac{(2\pi)^2\cos^2\theta}{\lambda^4\cos\theta_0}[4\cos^2\theta_0(\cos^2\phi\sin^2\zeta + \frac{\sin^2\phi\cos^2\zeta}{\cos^2\theta_0}$$

$$+4\cos^2\theta_0\left(\frac{(\cos\phi-\sin\theta_0\sin\theta)^2\cos^2\zeta}{\cos^2\theta_0\cos^2\theta}+\frac{\sin^2\phi\sin^2\zeta}{\cos^2\theta}\right)]g(\vec{K}), \tag{2}$$

where $\zeta=0$ for p polarization and $\zeta=\pi/2$ for s polarisation.

2.2 Scalar theory

Scalar theory does not take into account the polarizations of the incident and scattered fields nor the dielectric permittivity of the material. Never-the-less it is a simple solution which can be expected to give a qualitative picture of the angular distribution of the scattered field. The expression for the normalized scattered power is given by[2]

$$\frac{1}{P_0}\langle\frac{dP}{d\Omega}\rangle = \frac{16\pi^2}{\lambda^4}\frac{(\cos\theta_0+\cos\theta)^4}{\cos\theta}g(K), \tag{3}$$

with the same notations as earlier.

2.3 Power spectrum

All quantities in Eqs. (1)-(3), except the PSD function, can easily be measured or calculated to a sufficiently high degree of accuracy. The PSD function can be estimated using analytical expressions and fitting the parameters of these expressions to experimental data. This procedure requires that one has a good idea about the statistics of the surface roughness. This is generally not the case. We avoid this problem by using experimental surface profile data to calculate an estimate for the power spectrum (represented by the PSD function). Since we use one-dimensional surface profile data taken along a line on the surface we need to adapt Eqs. (1)-(3) to accommodate these data. This can be done using basically two different approximations. Either we assume that each realization of the rough surface, which is thought of as having been generated by a random process, is circularly concentric, or we assume that the surface roughness is statistically isotropic. For both these assumptions we have that $g(\vec{K})=g(|\vec{K}|)=g(K)$. For the circularly concentric model we find that[6]

$$g(K)=\lim_{R\to\infty}\frac{\langle|2\pi\int_0^R dr\,r\,\Delta z(r)J_0(Kr)|^2\rangle}{\pi R^2}, \tag{4}$$

Eq. (4) can be integrated numerically using experimental surface profile data $\Delta z(r)=\Delta z(x_i)$. R is taken to be equal to the measured profile length L. If L is large Eq. (4) is a good estimate for the power spectrum of the surface under consideration.

If the surface roughness is isotropic we can use the profile autocovariance function $C(\tau)$ to calculate an estimate for the isotropic two-dimensional power spectrum. $C(\tau)$ is related to the isotropic power spectrum via a zeroth-order Hankel transform, or[6]

$$g(K)=\frac{1}{2\pi}\int_0^\infty d\tau\,\tau J_0(K\tau)C(\tau)\,, \tag{5}$$

The autocovariance function can be calculated from experimental surface profile data using standard methods.[7] Eq. (5) can be integrated numerically for a very long surface profile to obtain an estimate for the PSD function.

In Fig. 1 we have plotted an averaged version of Eq. (5) (solid line) and the results for Eq. (4) for an aluminium surface AlSi4. Both the ensemble average in Eq. (4) and the averaging over the individual spectra from Eq. (5) was made for 35 200 µm profiles. The logarithm of the PSD function is plotted against the logarithm of the surface spatial wavenumber K. As we see both models yield results which agree well in the wavenumber domain contributing to the scattering in the near UV-visible wavelength region (indicated by the shaded area).

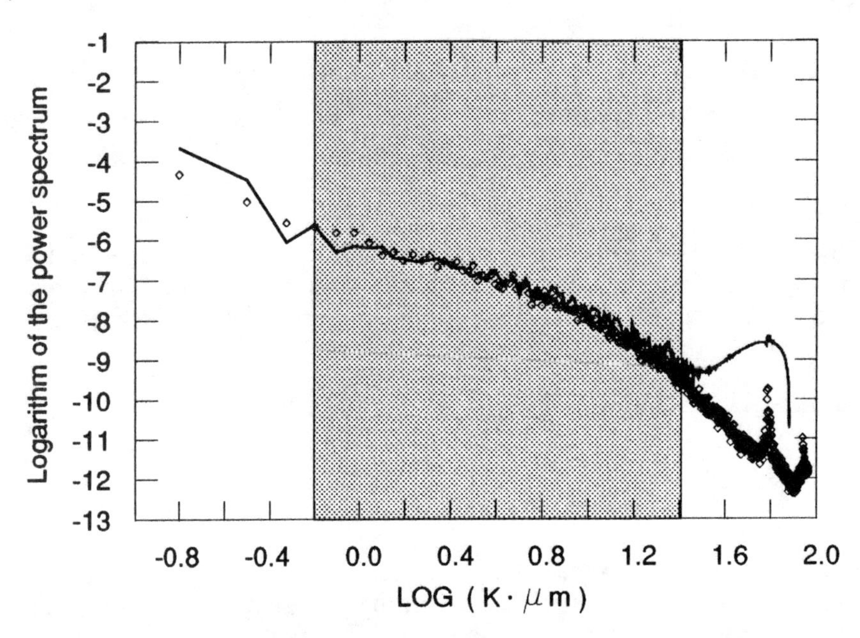

Figure 1. The power spectra for the isotropic (solid line) and the circularly concentric model (small squares). The bump at the end of the solid line is caused by numerical instabilities and the two peaks in the circularly concentric data are caused by instrumental noise.

2.4 Wavelength scaling

As apparent from Eqs. (1)-(3) the scattered power calculated to first order is proportional to λ^{-4}. For surfaces to which first-order solutions apply we should be able to observe, experimentally, this wavelength dependence. If we *measure* the scattering at wavelength λ_i and at wavelength λ_j and divide the measured values theory predicts that

$$\frac{\frac{1}{P_0}\left\langle\frac{dP}{d\Omega}\right\rangle_i}{\frac{1}{P_0}\left\langle\frac{dP}{d\Omega}\right\rangle_j}=\left(\frac{\lambda_j}{\lambda_i}\right)^4\frac{OPT(\varepsilon_1,\lambda_i,\theta,\theta_0)_i}{OPT(\varepsilon_1,\lambda_j,\theta,\theta_0)_j}\frac{g(K_i)}{g(K_j)}\,. \tag{6}$$

The relation between the optical factors, $OPT(\varepsilon_1,\lambda,\theta,\theta_0)$, can easily be calculated. For metals, like silver and aluminium, it is close to unity over a large angular domain. If we compare the measured data for the scattered power at different wavelengths for different angles θ_i and θ_j, such that $K_i=K_j$ ($K_i=(2\pi/\lambda_i)\sin\theta_i$), we expect to find that

$$\frac{\frac{1}{P_0}\langle\frac{dP}{d\Omega}\rangle_i}{\frac{1}{P_0}\langle\frac{dP}{d\Omega}\rangle_j}\frac{OPT(\varepsilon_1,\lambda_j,\theta,\theta_0)_j}{OPT(\varepsilon_1,\lambda_i,\theta,\theta_0)_i}=\left(\frac{\lambda_i}{\lambda_j}\right)^4=const. \quad . \tag{7}$$

This scaling property has been investigated experimentally both for two-dimensional and unidirectional surface roughness. For unidirectional surfaces the scattered intensity scales as λ^{-3}.

3. EXPERIMENTS

Theoretical calculations have been compared with experimental data for a number of different materials and surfaces, both isotropic and unidirectional (one-dimensional). Below we present the surfaces and the experimental procedures for the measurements.

3.1 Surface samples

In the experiments four types of surfaces were used, aluminium, copper, silver, and silicon. The silicon surfaces were taken from silicon wafers and etched, or polished, to various degrees of rms roughness. The metal surfaces were bulk surfaces which were diamond-turned and polished until the diamond-turning marks were removed.

All surfaces (except a set of silicon samples) were then polished by different procedures to obtain the required structure and degree of roughness. The surfaces used for the comparison with measured data for the angle-resolved scattering were polished to increasing levels of roughness. One of the goals of this study has been to determine the range of validity for first-order perturbation solutions. For the examination of the scaling properties of the scattered field we have prepared a set of surfaces of different materials. For each material the surface was polished to obtain an isotropic surface sample and a unidirectional sample. The idea behind this is that, even if we do not observe the predicted wavelength scaling, we should be able to see the difference in the scaling properties for unidirectional (one-dimensional) and isotropic (two-dimensional) surface roughness. We were careful to keep the roughness of these surfaces low in order to be in the domain of validity of first-order perturbation theory.

All surfaces were inspected visually in a Nomarski microscope for determination of a representative spot free from defects. The flatness of the surfaces was checked. The optical constants were taken from the literature.[8]

The surfaces have been characterized with respect to the root-mean-square (rms) roughness δ, to the ratio δ/λ, to $2\pi\delta/\lambda=k\delta$ and with respect to the lateral correlation length σ. Since the theoretical treatment in section 2 is based on the assumption that $\delta/\lambda<<1$ we investigate the agreement between theory and experiment as a function of δ/λ.

3.2 Surface profile measurements

All surface profiles were measured by a Talystep surface profiler.[9] The Talystep is often assumed to not distort the spectral contents of the surface roughness in the domain of physical interest for scattering in the UV-visible region. This has also been our assumption. The surfaces were measured in randomly oriented directions at the place on the surface where the scattering measurements were made. The profiles were compared with each other prior to any numerical treatment to check for and avoid possible anisotropy. The profile lengths varied from 200 μm to 1000 μm. All profiles were measured with a 0.3 μm radius stylus at a sampling distance of 0.034 μm. The number of profile height data in each sample ranged from 5868 to 29340 data points. The Talystep instrument that we have been using has a noise level of 0.4 Å rms and a non-profile-distorting lateral resolution of approximately 0.38 μm for the surfaces presented here.

3.3 Angle-resolved scattering measurements

The most important test of the validity of first-order perturbation theory has been to compare the numerical results of Eqs. (1) and (4) with experimental data for the angle-resolved scattering (ARS). The measurements were made for scattering in the plane of incidence for p-polarized laser beams. The experiments were performed at four different wavelengths using a 10 mW HeCd laser (λ=3250,4420 Å), a 2 mW HeNe laser (λ=6328 Å), and a diode laser (λ=8250 Å). The beam (Ø=1 mm) was incident on the surface at an angle of -11° to the surface normal. We used a collimator to ensure that the incident fields were as similar as possible for the different wavelengths. The scattering was measured in the range from 0° to 70° with respect to the specular beam. Because of interference with the specular beam in the instrument we have not compared theoretical and experimental results for the scattering at 0° and at 1°. The detector was a photomultiplier tube covering a solid angle of $3.0 \cdot 10^{-4}$ sr at 230 mm from the sample and it can be positioned with an accuracy of ±0.5° in the scattering hemisphere. The measured results have been corrected to give the normalized scattering per unit solid angle.

4. RESULTS

The results will be presented essentially in graphical form. The figures that are presented represent well the results that have been obtained. First we present the data for the comparison between theoretical calculations and experimental measurements for the angle-resolved scattered power. The results for the wavelength scaling are presented in section 4.2.

4.1 Angle-resolved scattering

In Fig. 2 we have plotted the measured (solid squares) and calculated data for the scattered power from an aluminium surface having an rms roughness of δ=87 Å. The wavelength of the incident field was λ=3250 Å (δ/λ=0.027). The lateral correlation length for this sample was 10.1 μm. The solid line represents the perturbation solution in Eq. (1), the dashed line is the infinitely conducting version of Eq. (1) (see Eq. (2)), and the crosses represent the scalar theory. The power spectrum was calculated using Eq. (4) for 35 different surface profiles. Care must be taken to assure that the proper definition for the Fourier transform is used for the power spectra. The definition of the PSD function may differ by a factor $(2\pi)^2$ between different perturbation-theory solutions. We see that both versions of the perturbation theory solution yield similar results and agree fairly well with the experimental results. They start to differ for large scattering angles. The scalar theory is not a very good approximation for this surface. We notice that both the angular distribution and the level of the scattered intensity disagrees with measured data for the scalar solution.

In Fig. 3 are presented the corresponding data for a silicon surface (δ=180 Å) for 3250 Å illumination. Silicon does not behave as an infinitely conducting surface as can be seen in Fig. 3. The infinitely conducting model, represented by the dashed line, yields results which differ from those from the full first-order perturbation solution in Eq. (1) (represented by the solid line). Four surface profiles were used for the ensemble average in Eq. (4). The agreement between theory and experiment is still rather good although, more surface profiles are needed for calculating the ensemble average in Eq. (4).

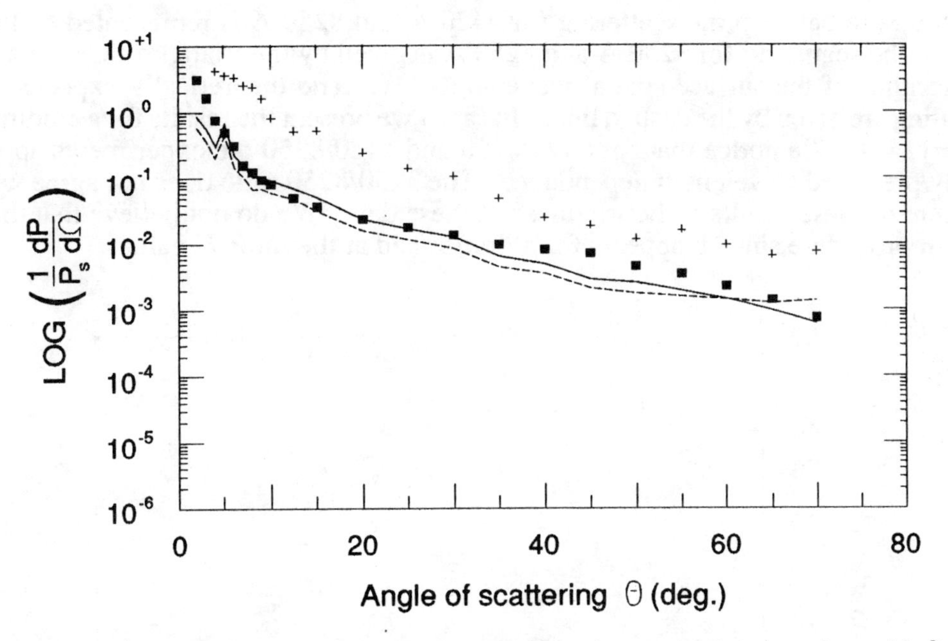

Figure 2. The calculated and measured (solid squares) scattered power for an aluminium surface with δ=87 Å. The two perturbation-theory solutions, represented by the dashed and the solid lines, yield agreement with measured data, whereas the scalar-theory solution (represented by the crosses) yields a less good agreement.

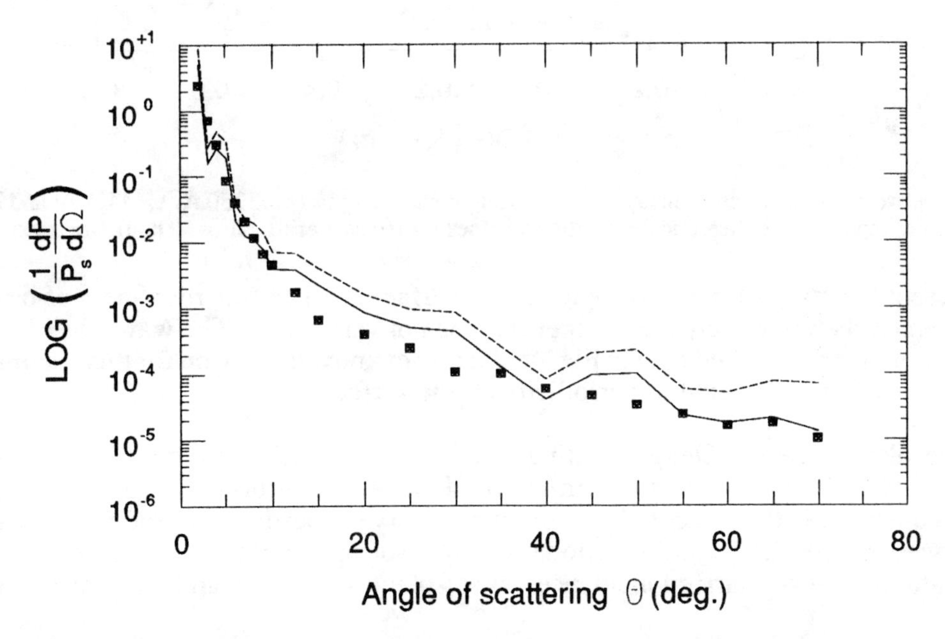

Figure 3. The measured and calculated normalized scattered power for a silicon surface etched to an rms roughness of 180 Å. This surface has a fractallike behaviour. As in Fig. 1 the solid line represents the solution in Eq. (1), and the infinitely conducting model is represented by the dashed line.

4.2 Wavelength scaling

The scattered power has been measured for four different wavelengths for both unidirectional and isotropic surface roughness. Theoretically the former has a λ^{-3} dependence, whereas the latter has a λ^{-4} dependence. In the figures below we present the ratios between the measured scattering at different wavelengths, according to the procedure described in section 2.4. The ratio of the scattered power for 6328 Å incidence and the scattered power for 8250 Å incidence (λ_i=6328 Å and λ_j=8250 Å in Eq. (7)) is represented by the solid squares, the ratio between the scattering for 4420 Å and 8250 Å is represented by the crosses, and the ratio between the scattering for 3250 Å and 8250 Å is given by the triangles. On the abcissa we have plotted the logarithm of the surface-spatial wavenumber *K*. The theoretically expected values for the wavelength scaling are given by the dashed lines. In Fig. 4 we present the results for a unidirectional copper surface with δ=114 Å. We notice that for 6328/8250 and 4420/8250 the experiments appear to confirm the theoretically predicted wavelength dependence. The 3250/8250 ratio does not agree with theory. An interesting feature of these results is the waviness of these data. We do not believe that this is caused by the measurement procedure since it appears for all ratios and at the same *K*-values.

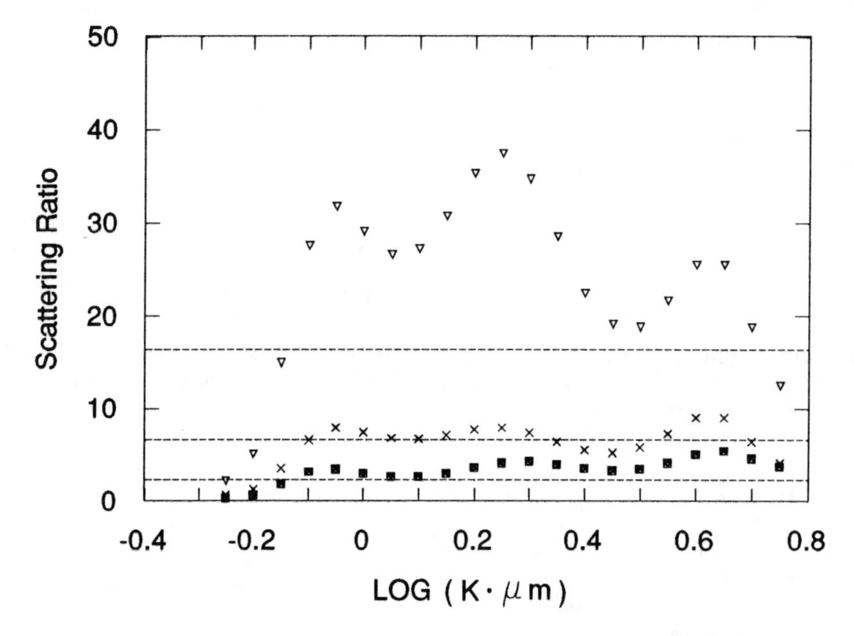

Figure 4. The ratio between the scattered intensity at three different wavelengths (λ_3=3250 Å, λ_4=4420 Å, and λ_8=8250 Å). The horizontal dashed lines represent the theoretically predicted values for the scattering ratios. The figure is explained in the text.

In Fig. 5 we present the results for an isotropic silicon surface with an rms roughness of δ=49 Å. For this surface the agreement between theory and experiment is not very good. The waviness of the ratios of the scattered power is more pronounced than in Fig. 4. We do not know the reason for this waviness but suspect that it might be caused by some kind of periodicity on the surface.

Generally, the results for the wavelength scaling were very difficult to interpret. We found surfaces for which the measured scaling properties confirmed the first-order perturbation theory solution as well as surfaces which contradicted the predicted perturbation-theory properties. We did not find any systematic differences between the results for unidirectional and for isotropic surfaces. An extended study of these phenomena would include s-polarized incidence, out-of-plane scattering, and more different surfaces.

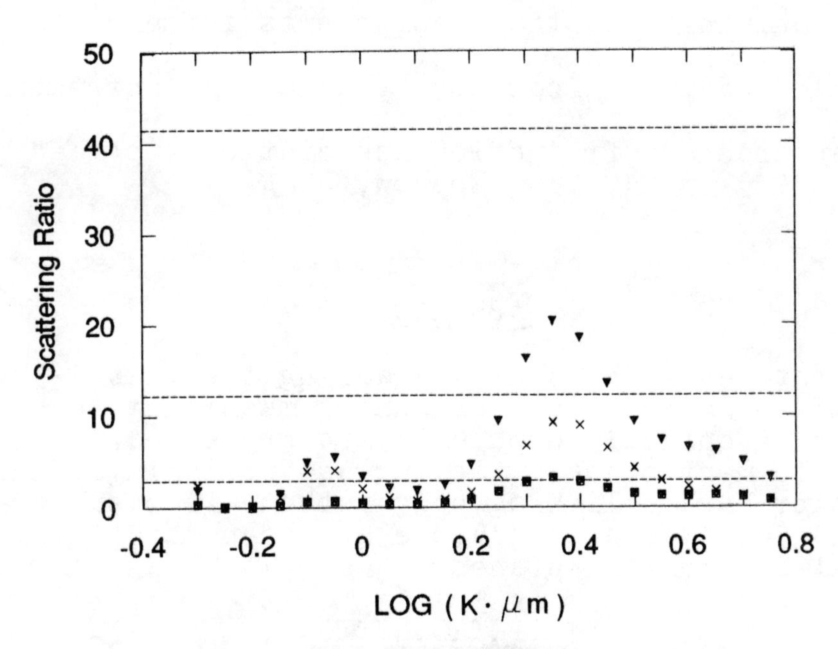

Figure 5. The ratios between the measured scattering for three wavelengths. The surface is an isotropic silicon surface. The agreement of the measured scattering data with the predicted λ^{-4} dependence is not so good. The figure is explained in the text.

5. CONCLUSIONS

On the basis of our results we conclude the following:
i) The investigated first-order perturbation theories yielded similar results for metal surfaces. For dielectric surfaces it is worth wile to use the full form of the first-order perturbation solution in Eq. (1).
ii) First-order perturbation theory can be used to predict the scattering from surface roughness with fairly good accuracy using experimental surface profile data if the ratio $\delta/\lambda<0.07$. The results appear to hold for different materials and wavelengths. This result has been confirmed earlier for one-dimensional surfaces in comparisons between first-order solutions and more rigourous integral solutions to the scattering problem.[10,11,12]
iii) We have not been able to experimentally verify the theoretically predicted wavelength dependence of the scattered intensity.

Notice that conclusions ii) and iii) are in contradiction to each other. It must be kept in mind, however, that the wavelength scaling test is a very sensitive test of the validity of perturbation theory.

6 REFERENCES

1 A. Ishimaru, *Wave Propagation and Scattering in Random Media* (Academic Press Inc., 1978).
2 H. Davies, Proc. IEE, **101**, 209(1954)
3 P. Beckmann and A. Spizzichino, *The Scattering of Electromagnetic Waves from Rough Surfaces* (Pergamon Press Ltd., 1963).
4 D. R. Jackson, D. P. Winebrenner, and A. Ishimaru, J. Acoust. Soc. Am., **83**, 961(1988).
5 J. M. Elson, Phys. Rev. B, **30**, 5460(1984).
6 E. L. Church, H. A. Jenkinson, and J. M. Zavada, Opt. Eng., **18**, 125(1979).
7 J. S. Bendat and A. G. Piersol, *Measurement and analysis of random data* (John Wiley & Sons., 1966).
8 E. D. Palik, *Handbook of Optical Constants of Solids*, (Academic Press Inc. London 1985).
9 J. M. Bennett and J. H. Dancy, Appl. Opt., **20**, 1785(1981).
10 M. F. Chen, S. C. Wu, and A. K. Fung, J. Wave Mater., **2**, 9(1978);
11 M. F. Chen and A. K. Fung, Radio Sci., **23**, 163(1988).
12 E. I. Thorsos and D. R. Jackson, J. Acoust. Soc. Am., **86**, 261(1989).

Finish machining of optical components in mass production

Alexander I. Grodnikov, Vladimir P. Korovkin

Research and Production Association "Optica"
103030, Moscow, USSR

ABSTRACT

Results that have been achieved during the last 3-5 years by the Soviet specialists in the field of finish machining technology in mass production of optical components are presented. It is shown that due to the development of new grinding and polishing means (a range of diamond tools, hard polishing polyurethane-foam substrates, bound abrasive tools)a considerable advance has been made in finish machining of optical spheres and planes.

1. INTRODUCTION

Soviet optical industry is characterized by commercial production of a good proportion of optical components, the quantity achieving tens and hundreds thousands a year. The ever increasing requirements for geometry precision and for quality of optically active surfaces of spherical and plane components used in binoculars, cine and photo technique, micrescopes, etc., state new and more complicated problems for the development of equipment, technology and tools for these purposes.

In overwhelming majority of production processes of optical glass machining loose abrasive yielded long ago to diamond tools, and for the most labour-intensive production operation, polishing, synthetic polishers on the basis of polyurethane foam are wider and wider used instead of polishing resins. High operating abilities of the diamond grinding and polyurethane-foam polishing tools and possibility to employ high-rate grinding and polishing operating conditions allow to reduce machining time at least by a factor of 10. However, due to a relative expensiveness of the diamond grinding and polyurethane-foam polishing tools, which are specialized for machining surfaces with a particular radius of curvature, their maximum efficiency is achieved when sufficiently large batches of components of the same type are produced.

Dozens types of optical glasses physical and mechanical properties of which differ by several times are commercially produced at optical plants. Characteristics of the processed surfaces (geometry accuracy, local defects, depth of the damaged layer) are defined mainly during fine diamond grinding and polishing.

To ensure at most perfect glass surfaces in combination with high-rate and stable processes of finish machining companies which specialize in making grinding and polishing means offer a wide choice of diamond pellets, polishing substrates, powders, pastes and resins.

The task of a process engineer who develops the process of optical surface machining is to find optimum number of technological iterations of fine grinding and polishing as well as to define required parameters of the tools.

Flow charts of standard processes of optical elements machining used in mass production in the Soviet optical industry are presented below with allowance for geometric features of surfaces, requirements to machining precision, physical and chemical properties of glass. It shoud be noted that all large plants, manufacturing medium- and high-precision optics, employ home-produced diamond tools and polishing means.

2. MACHINING OF PLANES AND BLOCKS OF SPHERICAL COMPONENTS WITH A LARGE RADIUS OF CURVATURE

Complete cycle of machining of single or blocked optical workpieces comprises rough grinding by ring diamond cutters using spheric or plane grinding machines, fine diamond grinding in several stages and polishing using machines without a force feed of the tool. Since the first operation presents no considerable problem and is standard actually for all types of optical components, we exclude it from further considerations.

Stages of finish machining of planes, including prisms and lenses with a large radius of curvature, are realized according to the flow chart presented in Figure 1. The main task in this case is to ensure high stability of the fine grinding process. The task is performed with increased-wear-rate diamond tools on a polymer binder.

As the tool wears out its required configiration is easily maintained due to a calculated distribution of the diamond-bearing layer over its working surface. Two-iteration polishing allows during a sufficiently short period of time (1-2 min) to remove the damaged layer using a rigid polisher and to ensure the surface shape accuracy $N = 1\text{-}2$ interference rings of an area 25x15 mm during the following 4-5 minutes of resin polishing.

For production of lenses with an aperture angle of $\leq 90°$ and shape accuracy $N = 1.0$ and $\Delta N = 0.2$, the diameter being up to 100 mm, we use the known technology named "Synchrospeed" (Fig. 2).

In the given flowcharts particular characteristics of diamond pellets, polyurethane foam, polishing powders and resins depend on the composition, on physical, mechanical and chemical properties of the glass being processed (crownglasses, flintglasses, borolanthanum glasses, etc.).

Soviet industry offers a wide range of diamond pellets for fine grinding, several types of polyurethane foam, resins and polishing compositions on the base of cerium and zirconium dioxides, thus promoting optimized choice of grinding and polishing means for particular types of optical articles.

3. MACHINING OF SINGLE AND BLOCKED SPHERICAL COMPONENTS WITH A SMALL RADIUS OF CURVATURE

The initial period of machining of single lenses and blocks of lenses close to hemispheres is characterized by bringing about high pressures on the tool's edge, this being due to a relatively large difference between the radia of the workpiece and of the finished optical piece (Fig. 3). The presented rational scheme of the tolerance removal excludes the employment of diamond pellets with a polymer binder in the second iteration of fine grinding due to a high wear rate of the tool's edge area independent of the filling factor of the working surface of the tool and, consequently, due to the fast loss of the specified geometry.In this case in the second iteration employment of tools with a metal binder is advantageous. A standard flow chart of finish machining of close to hemispheres surfaces with the diameter of up to 60 mm, made of glass type BK7 to a shape accuracy of 2-3 interference rings is given in Fig. 4. Finish machining with resin polishers is not needed provided that the requirements for surface accuracy don't exceed 4-5 interference rings.

However, for machining optical pieces made of soft glasses (flintglasses, for example)the proposed flow chart doesn't give positive results because of the poor quality of the surface after processing using a tool with a metal binder. A combined three-iteration fine grinding process is employed to overcome the problem (Fig. 5). A more uniform load distribution over the tool, employment of metal--polymeric and polymeric compositions promote the required stability of the surface shape and of the machining process as well as sufficient quality of the ground surface before polishing.

4. NEW PROCESS OF OPTICAL GLASS MACHINING WITHOUT APPLICATION OF LOOSE ABRASIVE

We have taken an advanced step in optical engineering - a production process for finish machining of optical workpieces which completely excludes application of loose abrasive in all iterations has been developed and brought into quantity production. The development was preceded by a detailed analysis of available from literature investigation results concerning the glass polishing mechanism and testing of polishing tool "Cerpet" of firm Fujimi (Japan) for operating ability.

The flow chart of the process comprises: fine diamond grinding in two iterations using a tool with a metal binder of grain size 24 and 12 mcm, superfine grinding using a tool with a polymer binder and polishing with the tool "Aquapol" with bound abrasive on the basis of cerium- or zircinium dioxide with content 90-92% (in mass) (Fig. 6).

"Aquapol" can be efficiently applied for surface finishing instead of optical resins since its geometry doesn't undergo changes in the real temperature range.

Investigations of contact interaction while polishing the glass using a tool with bound abrasive established a fundamental condition of its working ability, i.e. constant values of glass removal rate $\frac{dh_c}{d\tau}$ and of tool wear (Figure 7, curve1), with the exception of the initial stage of the process unlike those cases when a tool for fine diamond grinding is applied (Figure 7, curve 2).

The necessity to retain functional activity for an infinitely long time determines main requirements to a bound abrasive polishing tool and its operating conditions. Another important parameter determining the working ability of the tool is its wear, the relative value of which must not exceed 100%.

Depending on areas of application, the tool is available as cylindrical elements (pellets) with the diameter of 6, 8, 10, 15 and 20 mm and lapping spheres with the diameter of up to 50 mm. The central hole specially designed proceeding from the requirements of optimum filling curve allows to retain without dressing the tool's shape close to hemisphere (angle of cone $\gamma \rightarrow 90°$) practically to its wear out.

Optimal conditions and modes of finish machining of optical parts without application of loose abrasive are given in the table:

Indicators	Fine grinding		Superfine grinding	Polishing
	First iteration	Second iteration		
Tool rotational speed, s^{-1}	15-20	15-20	15	7-12
Unit pressure, 10^4 Pa	10-13	10-13	6-8	3-5
Filling factor	0.25	0.25	0.5	0.5-0.7
Initial roughness of the surface Ra, mcm	2.5-1.5	0.3-0.4	0.15-0.20	0.02-0.03
Allowance on the radius, mcm	4-5	2-3	1-2	0.5
Cooling agent	lubricant-coolant	lubricant-coolant	lubricant-coolant	water

Fields of effective applications are manufacturing of lenses, parts of microoptics, prisms, plates, final shaping of large optical parts using a small-area tool.

Indicators of the tool serviceability:

polishing ability, mcm/min .... not less than 0.5
tool relative wear, % .... not exceeding 100
accuracy of the machined
surface, interference rings .... $N \leqslant 2$; $\Delta N \leqslant 0.2$
roughness of the machined
surface Rz, mcm .... not exceeding 0.05

The developed technological process and tools ensure high-rate machining conditions, saving of ancillary materials and polishing powder, reduction in labour consumption while manufacturing optical parts, they improve working conditions in optical shops and create prerequisitesfor completely automated machining.

The finish machining process of optical surfaces without application of loose abrasive has been introduced into serial production at a number of optical plants.

5. CONCLUSION

Considerable results have been achieved in the development of a wide range of highly efficient tools and materials for optical glass machining. Their combined application allows to achieve high-rate and stable grinding and polishing processes. The designed polishing tool with bound abrasive is a new step in this direction. Thus, all prerequisites have been created for complete automatization of production processes of optical workpieces' machining and this, in its turn, will provide considerable advance in the production of mass optics.

6. ACKNOWLEDGEMENTS

The authors are grateful to all their colleagues for help in investigations and developments.

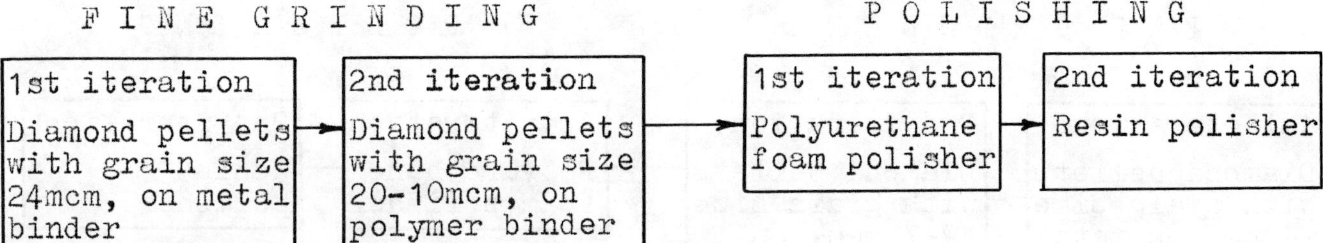

Figure 1. Flow chart of finish machining of planes and blocks of spherical components with a large radius of curvature

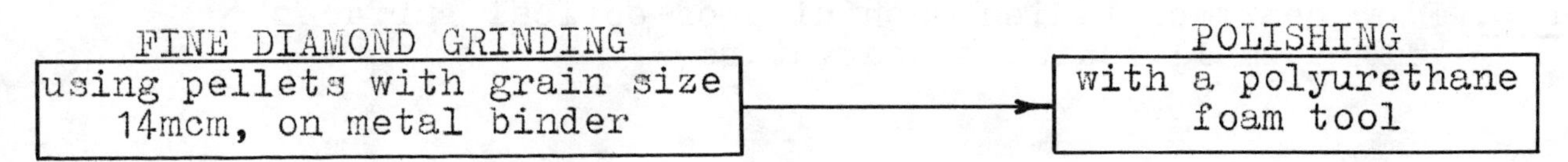

Figure 2. Flow chart of finish machining of lenses with an aperture angle ≤ 90° using "Synchrospeed" process

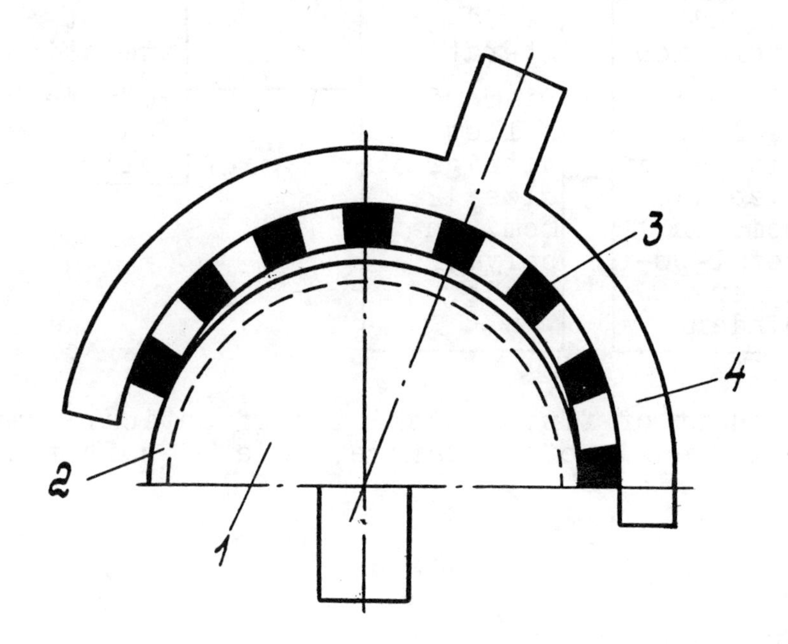

Figure 3. Scheme of stock removal while machining surfaces with a small radius of curvature:
1 - workpiece; 2 - stock allowed for machining;
3 - diamond pellets; 4 - tool body

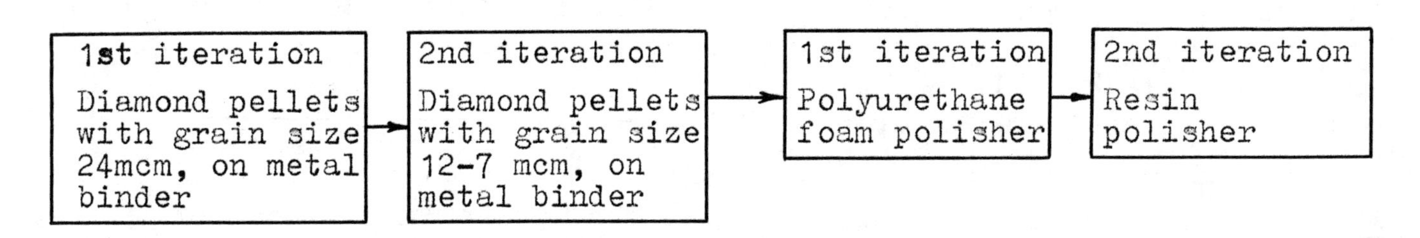

Figure 4. Flow chart of finish machining of optical surfaces with a small radius of curvature

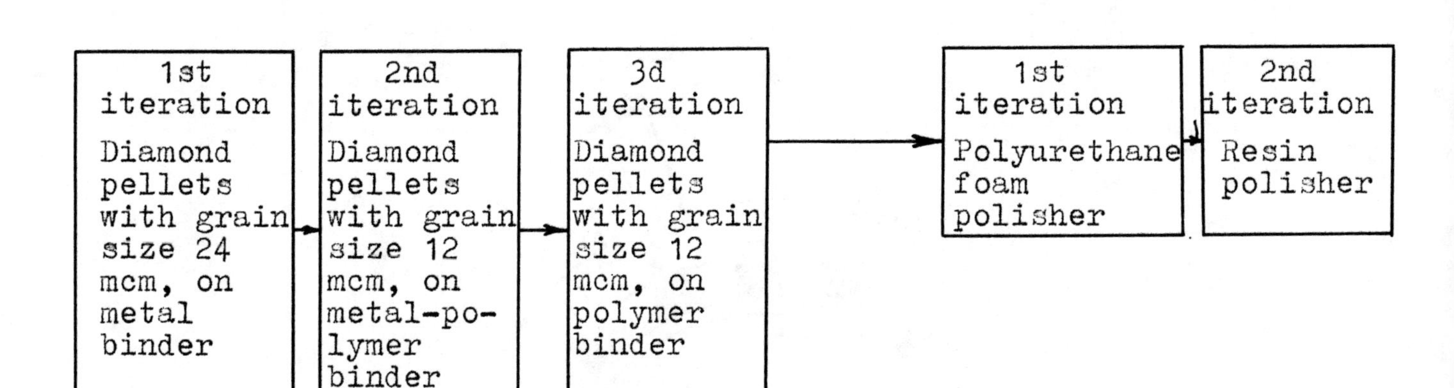

Figure 5. Flow chart of finish machining of optical components with a small radius of curvature, made of soft types of glass

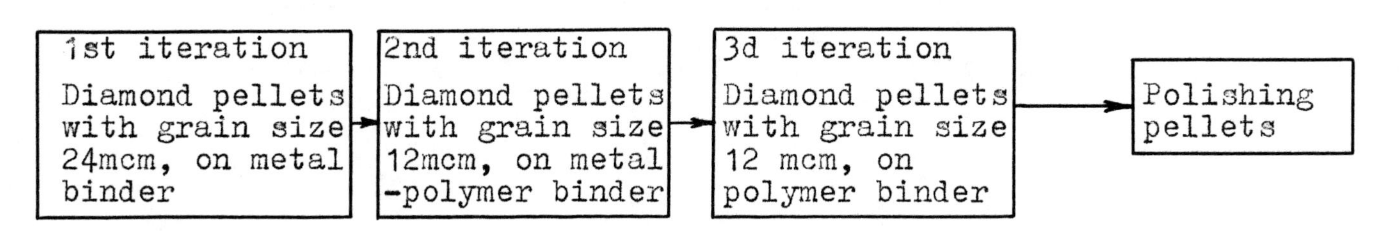

Figure 6. Flow chart of finish machining of optical components without application of loose abrasive

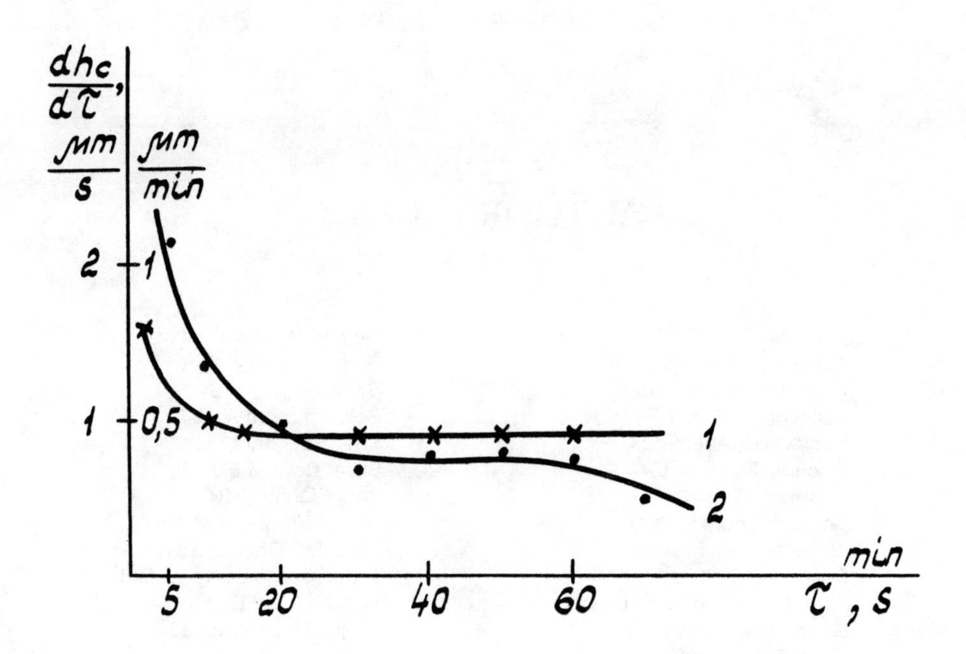

Figure 7. Glass removal rate variation:
1 - for polishing with the tool "Aquapol", mcm/min;
2 - for fine diamond grinding, mcm/s

AUTHOR INDEX